Broadband Planar Antennas

Broadband Planar Antennas

Design and Applications

Zhi Ning Chen and Michael Y. W. Chia

Both of
Institute for Infocomm Research, Singapore

John Wiley & Sons, Ltd

Copyright © 2006   John Wiley & Sons Ltd, The Atrium, Southern Gate, Chichester,
West Sussex PO19 8SQ, England

Telephone   (+44) 1243 779777

Email (for orders and customer service enquiries): cs-books@wiley.co.uk
Visit our Home Page on www.wiley.com

Other Wiley Editorial Offices

John Wiley & Sons Inc., 111 River Street, Hoboken, NJ 07030, USA

Jossey-Bass, 989 Market Street, San Francisco, CA 94103-1741, USA

Wiley-VCH Verlag GmbH, Boschstr. 12, D-69469 Weinheim, Germany

John Wiley & Sons Australia Ltd, 42 McDougall Street, Milton, Queensland 4064, Australia

John Wiley & Sons (Asia) Pte Ltd, 2 Clementi Loop #02-01, Jin Xing Distripark, Singapore 129809

John Wiley & Sons Canada Ltd, 22 Worcester Road, Etobicoke, Ontario, Canada M9W 1L1

Wiley also publishes its books in a variety of electronic formats. Some content that appears in print may not be available in electronic books.

Library of Congress Cataloging in Publication Data

Chen, Zhi Ning.
    Broadband planar antennas : design and applications / Zhi Ning Chen and Michael Y. W. Chia.
      p.   cm.
    Includes bibliographical references and index.
    ISBN-13: 978-0-470-87174-4 (cloth : alk. paper)
    ISBN-10: 0-470-87174-1 (cloth : alk. paper)
    1. Antennas (Electronics)   2. Microstrip antennas.   3. Broadband communication systems—Equipment and supplies.
    I. Chia, Michael Y. W.   II. Title.
    TK7871.6.C46 2005
    621.384′135—dc22

                 2005026868

British Library Cataloguing in Publication Data

A catalogue record for this book is available from the British Library

ISBN-13 978-0-470-87174-4 (HB)
ISBN-10 0-470-87174-1 (HB)

Typeset in 10/12pt Times by Integra Software Services Pvt. Ltd, Pondicherry, India.
Printed and bound in Great Britain by Antony Rowe Ltd, Chippenham, Wiltshire.
This book is printed on acid-free paper responsibly manufactured from sustainable forestry
in which at least two trees are planted for each one used for paper production.

To my father Dr Liang Chen, and my mother Madam Rui Zhi Wang.

Also to my wife Lin Liu, as well as my twin sons Shi Feng and Shi Ya, with special appreciation.

Zhi Ning Chen

To my wife Stella, and my children John and Grace.

Michael Y. W. Chia

Contents

Foreword   ix

Preface   xi

Acknowledgements   xiii

1 Planar Radiators   **1**
  1.1 Introduction   1
  1.2 Bandwidth Definitions   2
      1.2.1 Impedance Bandwidth   3
      1.2.2 Pattern Bandwidth   3
      1.2.3 Polarization or Axial-ratio Bandwidth   4
      1.2.4 Summary   5
  1.3 Planar Antennas   5
      1.3.1 Suspended Plate Antennas   5
      1.3.2 Bent Plate Antennas   10
  1.4 Overview of this Book   14
  References   15

2 Broadband Microstrip Patch Antennas   **17**
  2.1 Introduction   17
  2.2 Important Features of Microstrip Patch Antennas   20
      2.2.1 Patch Shapes   20
      2.2.2 Substrates   20
      2.2.3 Feeding Structures   21
      2.2.4 Example: Rectangular Microstrip Patch Antennas   23
  2.3 Broadband Techniques   31
      2.3.1 Lowering the Q 31
      2.3.2 Using an Impedance Matching Network   32
      2.3.3 Case Study: Microstrip Patch Antenna with Impedance Matching Stub   34
      2.3.4 Introducing Multiple Resonances   37
      2.3.5 Case Study: Microstrip Patch Antenna with Stacked Elements   39
  References   43

3 Broadband Suspended Plate Antennas   **47**
   3.1 Introduction   47
   3.2 Techniques to Broaden Impedance Bandwidth   49
       3.2.1 Capacitive Load   49
       3.2.2 Slotted Plates   51
       3.2.3 Case Study: SPA with an Ω-shaped Slot   52
       3.2.4 Electromagnetic Coupling   56
       3.2.5 Nonplanar Plates   59
       3.2.6 Vertical Feed Sheet   62
   3.3 Techniques to Enhance Radiation Performance   65
       3.3.1 Radiation Characteristics of SPAs   66
       3.3.2 SPA with Dual Feed Probes   72
       3.3.3 Case Study: Center-concaved SPA with Dual Feed Probes   75
       3.3.4 SPA with Half-wavelength Probe-fed Strip   77
       3.3.5 SPA with Probe-fed Center Slot   81
       3.3.6 Case Study: Center-fed SPA with Double L-shaped Probes   92
       3.3.7 SPA with Slots and Shorting Strips   100
   3.4 Arrays with Suspended Plate Elements   111
       3.4.1 Mutual Coupling between Two Suspended Plate Elements   112
       3.4.2 Reduced-size Array above Double-tiered Ground Plane   117
   References   131

4 Planar Inverted-L/F Antennas   **135**
   4.1 Introduction   135
   4.2 The Inverted-L/F Antenna   137
   4.3 Broadband Planar Inverted-F/L Antenna   141
       4.3.1 Planar Inverted-F Antenna   141
       4.3.2 Planar Inverted-L Antenna   144
   4.4 Case Studies   154
       4.4.1 Handset Antennas   154
       4.4.2 Laptop Computer Antennas   171
   References   174

5 Planar Monopole Antennas and Ultra-wideband Applications   **179**
   5.1 Introduction   179
   5.2 Planar Monopole Antenna   181
       5.2.1 Planar Bi-conical Structure   181
       5.2.2 Planar Monopoles   181
       5.2.3 Roll Monopoles   183
       5.2.4 EMC Feeding Methods   192
   5.3 Planar Antennas for UWB Applications   193
       5.3.1 Ultra-wideband Technology   193
       5.3.2 Considerations for UWB Antennas and Source Pulses   195
       5.3.3 Planar UWB Antenna and Assessment   212
   5.4 Case Studies   218
       5.4.1 Planar UWB Antenna Printed on a PCB   220
       5.4.2 Planar UWB Antenna Embedded into a Laptop Computer   227
       5.4.3 Planar Directional UWB Antenna   232
   References   237

Index   **241**

Foreword

With the rapid growth of wireless communications, broadband planar antennas are in strong demand to cover various applications with fewer antennas. This book describes recent advances in broadband planar antennas for wireless phones and LANs. In addition, the book has a unique feature in that it covers the special design requirements for broadband planar antennas in UWB radio systems.

The authors, Dr Zhi Ning Chen and Dr Michael Y. W. Chia, have a lot of computational and experimental experience in the design of these antennas. Dr Chen studied broadband planar antennas at the University of Tsukuba from 1997 to 1999 as a JSPS (Japan Society for the Promotion of Science) Fellow and started to work at the Institute for Infocomm Research with Dr Chia in 1999. Since then we have collaborated closely on broadband antenna research. In 2003, Dr Chen obtained a Doctor of Engineering degree at the University of Tsukuba, the title of his dissertation being *Broadband Design of Monopoles in Two-Plate Waveguides and Suspended Plate Antennas*. Some of the content of his dissertation is included in this book, and I am greatly honored and happy to write this foreword.

Chapter 1 explains the fundamental antenna indices and the inherent relationships between different types of planar antenna. Chapter 2 covers important features and broadband techniques for microstrip patch antennas. Chapter 3 discusses broadband suspended plate antennas and arrays, which have been studied extensively by the authors. Techniques are described to get broadband impedance, to enhance radiation performance and to improve diversity performance. Chapter 4 discusses broadband planar inverted-L/F antennas and their application to handsets and laptop computers. Chapter 5 covers planar monopole antennas and their UWB applications. Antenna design considerations for UWB radio systems are discussed in the frequency and time domains.

I believe that this is the first comprehensive book on planar antennas that shows various broadband techniques to design these antennas for practical applications. It is well organized and has plentiful diagrams with numerical and experimental results to help understand the concepts. There are many references at the end of each chapter to assist with further study of each topic. I believe that this book will be of interest and useful to anyone who works on planar antennas, their broadband techniques and their application to UWB radio systems.

Professor Kazuhiro Hirasawa
Graduate School of Systems and Information Engineering
University of Tsukuba
Japan

Preface

We have been fortunate to witness and be part of the advances in communications technology over the last two decades. In particular, mobile communications which have drastically changed the lifestyles of people through the widespread use of wireless personal terminals for both voice and data services. Connecting people anytime and anywhere has become a reality, be it talking to friends on a mobile phone while on a train, or chatting with a stranger at an internet café. As antenna engineers, we are able to contribute to this wireless revolution by designing high performance antennas for fixed base-stations and mobile portable terminals. The demands placed on mobile communication systems have increased remarkably, so antenna designers face many challenges: small or tiny size, low- or ultra-low-profile features, wide- or ultra-wide operating bandwidth, multiple functions, pure polarization, low cost, etc. As a result, antenna research and development has become a hot topic in both academia and industry, as demonstrated by the increasing number of publications and start-up companies.

One of the authors (Zhi Ning Chen) started his research into antennas for mobile communication systems at the Centre for Wireless Communication, Singapore, in 1999, with particular emphasis on broadband technology. Planar structures were selected as the broadband solution because of their lower profile, broader bandwidth, more design degrees, and ease of manufacture by virtue of the simpler geometry. They include antennas with planar radiators such as microstrip antennas, suspended plate antennas, planar inverted L/F antennas, and planar monopole antennas. Applications have covered cellular phone systems, global positioning systems, wireless local area networks, vehicle mobile communication systems, and ultra-wideband-based wireless personal area networks. The aim in writing this book was to introduce readers systematically to the latest progress in planar broadband antennas with the review of conventional bandwidth techniques, information on which is scattered in countless journals, conference proceedings and books. All the results shown in this book have been verified by experiment or simulation or both. This book will therefore be an invaluable design aid for all 'ants' (antenna researchers, engineers and students).

Broadband Planar Antennas is organized into five chapters. Chapter 1 introduces the concepts, with a brief look at the radiation characteristics of planar radiators with different heights. Roughly, planar radiators can be categorized into (a) microstrip patch antennas with very thin dielectric substrates, (b) suspended plate antennas with thick and low-permittivity dielectric substrates, (c) planar monopole antennas, a variation of the thin-wire monopole, and (d) inverted-L/F antennas, situated between suspended plate antennas and planar monopole antennas.

Chapter 2 reviews techniques to broaden the bandwidth of a microstrip patch antenna. The methods include lowering the Q, using a matching network, and introducing multiple resonances.

Chapter 3 covers three aspects of suspended plate antennas (SPAs). First, broadband techniques based on the concept of low antenna Q are considered. With the various techniques, impedance bandwidths of the SPAs reach up to 60 %. Second, there is a discussion of techniques to enhance the radiation performance of broadband SPAs. Balanced excitation and symmetrical configurations are considered as the basic ideas to improve radiation performance. The final section of Chapter 3 describes design techniques for SAP arrays. The design goal is to suppress mutual couplings and reduce the lateral size of the broadband arrays with suspended elements.

Chapter 4 looks at techniques for broadening the impedance bandwidth of planar inverted-L/F antennas (PIFAs), especially when used in portable devices such as handphones, PDAs and laptops. In particular, planar inverted-F antennas with a small system ground plane in handphones are described.

Chapter 5 covers broadband planar monopole antennas, and explores their application in emerging ultra-wideband (UWB) systems.

Zhi Ning Chen
Michael Y. W. Chia
Singapore

Acknowledgements

This book was written at the invitation of Sarah Hinton of John Wiley & Sons, Ltd. We appreciate her patience while we completed the task. We also acknowledge the kind assistance of ex-students, Terence See, Xuan Hui Wu, Ning Yang, Yan Zhang and Hui Feng Li. Terence, in particular, read through the whole book and provided many helpful comments. In addition, we would like to give thanks to many friends and colleagues who have generously offered us their encouragement.

Finally, we owe much to our wives and children for their love, tolerance and patience during our prolonged work on this book. The time missed together during our research and writing can never be replaced.

Zhi Ning Chen
Michael Y. W. Chia
Singapore

1

Planar Radiators

1.1 INTRODUCTION

The rapid development of wireless communication systems is bringing about a wave of new wireless devices and systems to meet the demands of multimedia applications. Multi-frequency and multi-mode devices such as cellular phones, wireless local area networks (WLANs) and wireless personal area networks (WPANs) place several demands on the antennas. Primarily, the antennas need to have high gain, small physical size, broad bandwidth, versatility, embedded installation, etc. In particular, as we shall see, the bandwidths for impedance, polarization or axial ratio, radiation patterns and gain are becoming the most important factors that affect the application of antennas in contemporary and future wireless communication systems.

Table 1.1 shows the operating frequencies of some of the most commonly used wireless communication systems. The bandwidths vary from 7 % to 13 % for commercial mobile communication systems, and reach up to 109 % for ultra-wideband communications. The antennas used must have the required performance over the relevant operating frequency range. Antennas for fixed applications such as cellular base-stations and wireless access points should have high gain and stable radiation coverage over the operating range. Antennas for portable devices such as handphones, personal digital assistants (PDAs) and laptop computers should be embedded, efficient in radiation and omnidirectional in coverage. Most importantly, the antennas should be well impedance-matched over the operating frequency range. For example, an array designed for a cellular base-station operating in the GSM1900 band should have an impedance bandwidth of 7.3 % for a return loss of less than −15 dB. Antennas for mobile terminals must be small in physical size so that they can be embedded in devices or conform to device platforms. More often than not, the antennas are electrically small in size, which significantly narrows the impedance bandwidth and greatly reduces radiation efficiency or gain.

For base-stations, antennas or arrays must be compact to reduce installation costs and to harmonize aesthetically with the environment, but the reduced size generally results

Broadband Planar Antennas: Design and Applications   Zhi Ning Chen and Michael Y. W. Chia
© 2006 John Wiley & Sons, Ltd

Table 1.1 Wireless communication system frequencies.

System	Operating frequency	Overall bandwidth
Advanced Mobile Phone Service (AMPS)	Tx: 824–849 MHz Rx: 869–894 MHz	70 MHz (8.1 %)
Global System for Mobile Communications (GSM)	Tx: 880–915 MHz Rx: 925–960 MHz	80 MHz (8.7 %)
Personal Communications Service (PCS)	Tx: 1710–1785 MHz Rx: 1805–1880 MHz	170 MHz (9.5 %)
Global System for Mobile Communications (GSM)	Tx: 1850–1910 MHz Rx: 1930–1990 MHz	140 MHz (7.3 %)
Wideband Code Division Multiple Access (WCDMA)	Tx: 1920–1980 MHz Rx: 2110–2170 MHz	250 MHz (12.2 %)
Universal Mobile Telecommunication Systems (UMTS)	Tx: 1920–1980 MHz Rx: 2110–2170 MHz	250 MHz (10.2 %)
Ultra-wideband (UWB) for communications and measurement	3100–10 600 MHz EIRP: < -41.3 dBm	7500 MHz (109 %)

in a degraded radiation performance. Moreover, varying installation environments require antennas with a big bandwidth tolerance. Consequently, the bandwidth requirement for small or compact antennas has become a very critical design issue. Researchers in academia and industry have devoted much effort to the development of a variety of techniques for such small or compact broadband antenna designs. Antennas with broad bandwidths have additional advantages, such as to mitigate design and fabrication tolerances, to reduce impairment due to the installation environment, and most importantly, to cover several operating bands for multi-frequency or multi-mode operations.

1.2 BANDWIDTH DEFINITIONS

The bandwidth of an antenna may be defined in terms of one or more physical parameters. As shown in equation 1.1, the bandwidth may be calculated by using the frequencies f_u and f_l at the upper and lower edges of the achieved bandwidth:

$$
\text{BW} = \begin{cases} \dfrac{2\,(f_u - f_l)}{f_u + f_l} \times 100\,\% & \text{bandwidth} < 100\,\% \\ \dfrac{f_u}{f_l} : 1 & \text{bandwidth} \ge 100\,\%. \end{cases}
\tag{1.1}
$$

The bandwidth of an antenna can be defined for *impedance, radiation pattern* and *polarization*. First, a satisfactory impedance bandwidth is the basic consideration for all antenna design, which allows most of the energy to be transmitted to an antenna from a feed or a transmission system at a transmitter, and from an antenna to its load at a receiver in a wireless communication system. Second, a designated radiation pattern ensures that maximum or minimum energy is radiated in a specific direction. Finally, a defined polarization of an antenna minimizes possible losses due to polarization mismatch within its operating bandwidth. The bandwidths used in this book will be targeted at the impedance, radiation pattern and polarization bandwidths, which are defined below.

1.2.1 IMPEDANCE BANDWIDTH

In general, an antenna is a resonant device. Its input impedance varies greatly with frequency even though the inherent impedance of its feed remains unchanged. If the antenna can be well matched to its feed across a certain frequency range, that frequency range is defined as its *impedance bandwidth*. The impedance bandwidth can be specified in terms of *return loss* (S parameter: $|S_{11}|$) or a *voltage standing-wave ratio* (VSWR) over a frequency range. The well-matched impedance bandwidth must totally cover the required operating frequency range for some specified level, such as VSWR = 2 or 1.5 or a return loss $|S_{11}|$ of less than $-10\,\text{dB}$ or $-15\,\text{dB}$. Furthermore, the impedance bandwidth is inversely proportional to the *quality factor* (Q) of an antenna as given by

$$\text{BW} = \frac{\text{VSWR} - 1}{Q\sqrt{\text{VSWR}}}. \tag{1.2}$$

According to Chu's criterion, the minimum quality factor $Q_{\min}$ of an antenna of a given size is given approximately by

$$Q_{\min} = \frac{1 + 3(k_0R)^2}{(k_0R)^3[1 + (k_0R)^2]}. \tag{1.3}$$

The theory states that the radius R of the minimum sphere that completely encloses the antenna determines the $Q_{\min}$ of an antenna with 100 % radiation efficiency. Chu's criterion predicts the maximum impedance bandwidth of the antenna. However, an antenna usually has a much higher Q than the $Q_{\min}$ given in equation 1.3. For example, a microstrip antenna with a thin and high-permittivity dielectric enclosed by a sphere with a large radius R suffers from a narrow impedance bandwidth (typically a few percent) due to its high Q. Other antennas, such as wire monopoles, enclosed by the same sphere may have much broader bandwidths than the microstrip antenna. Therefore, taking advantage of the full volume of the enclosing sphere may enhance the impedance bandwidth.

Others have greatly extended Chu's work.[1,2,3] Recently, this criterion has been challenged by a possible smaller or even zero Q.[4] On the other hand, a lower Q indicates not only a broader bandwidth but also a higher loss. For a resonator, the lower Q can be caused by larger losses such as leaky energy and ohmic losses, which are definitely undesirable for an antenna. Therefore, it is necessary to examine radiation and antenna efficiency or gain (defined as the antenna directivity multiplied by antenna efficiency) across the whole impedance bandwidth.

1.2.2 PATTERN BANDWIDTH

There are many parameters to describe the radiation performance of an antenna, including the following:

- main beam direction
- side-lobe, back-lobe, grating-lobe levels and directions
- beamwidth and half-power beamwidth
- beam coverage and beam solid angle
- front-to-back ratio
- directivity
- efficiency

- phase center
- gain and realized gain
- effective area, effective height and polarization.

These all vary with frequency, and the operating frequency range can be determined by specifying any of these parameters as either a minimum or a maximum according to the system requirements for the antenna. Variations in the parameters result essentially from frequency-dependent distributions of the magnitudes and phases of electric and magnetic currents on antenna surfaces.

Radiation patterns are important indicators of the operating modes of an antenna. Usually the radiation patterns are expressed in Cartesian or polar coordinate systems. The Cartesian system is used in this book, as depicted in Figure 1.1.

1.2.3 POLARIZATION OR AXIAL-RATIO BANDWIDTH

The parameters relating to polarization characteristics are important. Besides the parameters mentioned above, additional parameters include the axial ratio, the tilt angle and the sense of rotation. The polarization properties of a linearly or circularly polarized antenna should be specified to avoid losses due to polarization mismatch.[5] The bandwidth can be defined by specifying a maximum *cross-polarization* (or *cross-pol*) level or *axial-ratio* level. The bandwidth should entirely cover the operating frequency range.

Control of polarization depends on the control of orthogonal modes excited in linear and circular antennas. The isolation between the orthogonal modes determines the cross-pol level or axial-ratio level. An antenna's Q and excitation greatly affect this isolation. Generally, a low Q or a broad impedance bandwidth results in poor isolation between the orthogonal modes. Therefore, it is difficult to enhance both impedance and polarization bandwidths simultaneously by reducing the antenna's Q. One possible way to enhance the polarization performance within a broad bandwidth is to carefully design the excitation geometry. In particular, the ratio of the maximum co-polarization to cross-polarization radiation levels

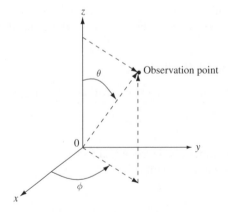

Figure 1.1 The Cartesian coordinate system as used in this book.

(the *co-to-cross-pol ratio*) is used to evaluate the polarization purity. Within the beamwidth of interest, the co-to-cross-pol ratio can be calculated from

$$\text{Co-to-cross-pol ratio} = \frac{\text{maximum co-polarization radiation level}}{\text{maximum cross-polarization radiation level}}. \tag{1.4}$$

1.2.4 SUMMARY

In short, the accomplishment of acceptable bandwidths is definitely an important and critical consideration for antenna design in wireless communication systems. A well-matched impedance bandwidth must cover the entire required operating frequency range for some specified levels, such as VSWR = 2 or 1.5 or a return loss $|S_{11}|$ less than -10 dB or -15 dB. Next, over the required bandwidth, the gain, beamwidth and the radiation patterns of an antenna must be stable to meet the system requirements. Finally, across the operating bandwidth, the co-to-cross-pol ratios in specified plane cuts must be higher than the delimited values for applications requiring polarization purity.

This book will review a variety of the popular broadband techniques that have been developed by various researchers in the past decades, and highlight the latest advances in broadband antenna design. The bandwidths in terms of both impedance and radiation performance will be discussed. Other design considerations such as antenna size and shape, materials and fabrication cost as well as complexity are also taken into account.

1.3 PLANAR ANTENNAS

In general, all antennas comprising planar or curved surface radiators or their variations and at least one feed are termed 'planar antennas'. Printed microstrip patch antennas, slot antennas, suspended plate antennas, planar inverted-L and inverted-F antennas (PILAs and PIFAs), sheet monopoles and dipoles, roll monopoles, and so on, are typical planar antennas used extensively in wireless communication systems. Usually, they exhibit merits such as simple structure, low cost, low profile, small size, high polarization purity or broad bandwidth. The following subsections discuss the impedance and radiation characteristics of two simple but typical planar antennas. This will serve to demonstrate the inherent relationship between the different planar antennas.

1.3.1 SUSPENDED PLATE ANTENNAS

Figure 1.2 shows the geometry of a planar antenna with a square, perfectly electrically conducting (PEC) radiator. The radiating PEC plate, measuring $l \times l = 70$ mm $\times$ 70 mm, is suspended parallel with a ground plane (x–y plane); h denotes the spacing between the radiating plate and the ground plane. A cylindrical 50-Ω probe having a 0.6 mm radius excites the radiating plate through the ground plane. The feed point is a distance d from the edge of the plate and positioned in the midline of the radiator along the y-axis. By adjusting the position of the feed point, the antenna with different spacing h can achieve good impedance matching between the plate and the feeding probe.

Figure 1.3 illustrates the variation in the input impedance against frequency for varying h. Within the frequency range 1.5–2.5 GHz, the impedance loci are plotted in a counter-clockwise

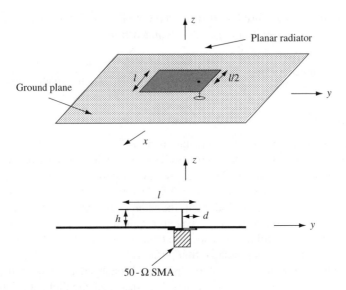

Figure 1.2 A square plate antenna with size $l \times l$ suspended above a ground plane with spacing h between the ground and radiator, and fed by a cylindrical probe.

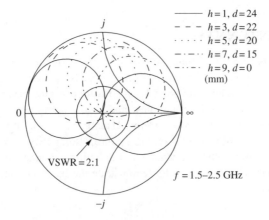

Figure 1.3 Input impedance versus frequency of a square planar antenna.

manner. The loops of the impedance loci gradually move from a low-impedance region (the left-hand side of a Smith chart) to a high-impedance region (the right-hand side) when the spacing h increases from 1 mm to 9 mm. In order to achieve good impedance matching, the feed point has shifted from $d = 24$ mm to $d = 0$ mm (the edge of the plate).

The return losses $|S_{11}|$ against frequency are shown in Figure 1.4. It is evident that an antenna with a spacing h of 1 mm has the highest Q and lowest radiation efficiency of about 85 % at its resonant frequency. With increasing h, the Q of the antenna significantly decreases, and the radiation efficiency increases correspondingly up to 99 % as the spacing reaches 9 mm.

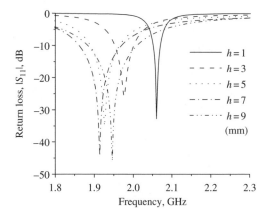

Figure 1.4 Return loss versus frequency for a square planar antenna.

The resonant frequencies for minimum return losses, and the achieved bandwidths for $|S_{11}|$ less than $-10\,\mathrm{dB}$, are shown in Figure 1.5. With increasing h, the impedance bandwidths increase from 1 % to 7 % due to the lower Q, and the resonant frequencies decrease from 2.07 GHz to 1.92 GHz due to the larger height of the antenna (except for the case when $h = 9\,\mathrm{mm}$).

The input impedance loci illustrated in Figure 1.6 show that further increasing the spacing h from 9 mm to 21 mm leads to a larger input impedance and poorer impedance matching. In particular, the longer probe causes a larger input inductance around the resonant frequency. Meanwhile, the feed point has reached the edge of the plate, so it is impossible to improve the impedance matching by further shifting the feed. Therefore, it is necessary to apply broadband impedance matching techniques for antennas with large spacing, as discussed in Chapter 3.

The radiation performance of the antenna will be examined on two principal planes, E (y–z) and H (x–z). Figures 1.7 and 1.8 illustrate the co- and cross-pol radiation patterns for the gain. The data were obtained at the corresponding resonant frequencies when the spacing h varies from 1 mm to 9 mm.

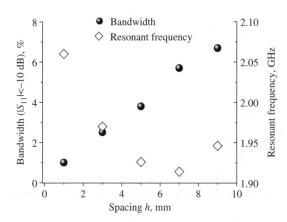

Figure 1.5 Achieved impedance bandwidth and resonant frequency versus spacing h.

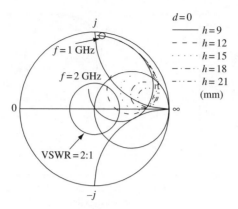

Figure 1.6 Input impedance versus frequency for $h = 9$ mm to 21 mm, $f = 1$–2 GHz.

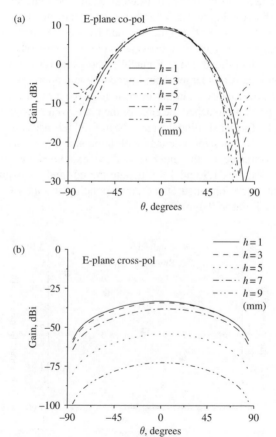

Figure 1.7 Radiation patterns in the E-plane for $h = 1$ mm to 9 mm.

Figure 1.7(a) shows the co-pol radiation patterns in the E-plane. The gain of the at antennas are from 9.1 dBi to 9.5 dBi the boresight ($\theta = 0°$, $\phi = 0°$) with half-power (3-dB) beamwidths of 57–60°. With decreasing resonant frequency for larger spacing h, the asymmetrical side lobes gradually appear due to the asymmetrical structure with respect to the x–z plane. The highest side-lobe level reaches 14 dB below the maximum major-lobe level when $h = 9$ mm.

Figure 1.7(b) shows the cross-pol radiation patterns in the E-plane. The co-to-cross-pol ratios in the boresight increase from 42 dB to 82 dB as the spacing h increases from 1 mm to 9 mm. Therefore, the cross-polarized radiation in the E-plane, is not severe compared with those in the H-plane, as shown in Figure 1.8.

Figure 1.8(a) demonstrates that the co-pol radiation patterns in the H-plane remain symmetrical with respect to the E-plane, with slight variations. Moreover, the gain and half-power beamwidths are essentially unchanged. Thus, compared with the co-pol radiation patterns in the E-plane, the co-pol radiation patterns in the H-plane are more stable.

The cross-polarized radiation in the H-plane depicted in Figure 1.8(b) becomes an important factor in determining the radiation performance of a planar antenna. Due to the symmetrical structure with respect to the y–z plane, the cross-pol radiation patterns in the H-plane are symmetrical as well. With increasing h, the co-to-cross-pol ratios significantly

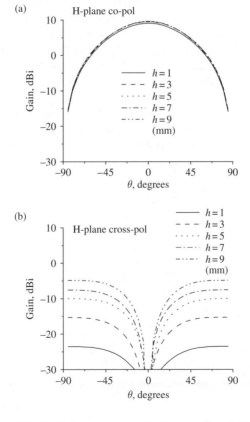

Figure 1.8   Radiation patterns in the H-plane ($h = 1$–9 mm).

decrease from 33 dB to 14.5 dB. Within the half-power beamwidths of the co-pol radiation patterns, the co-to-cross-pol ratios vary from 35 dB to 17 dB. For applications such as cellular base-station antenna arrays that require high polarization purity, these co-to-cross-pol ratios are too low, so techniques have been developed to alleviate the problem of high cross-pol radiation across a broad bandwidth, as discussed in Chapter 4.[6-12]

The variations of the co-to-cross-pol ratios across the impedance bandwidths will next be examined for antennas with $h = 1$ mm and 9 mm. The cross-pol radiation patterns at the lower and upper edge frequencies (f_l, f_u) of the bandwidths and their resonant frequencies f_r in the H-plane are compared in Figure 1.9. Evidently, the cross-pol radiation levels at f_u are the highest within the bandwidths for $h = 9$ mm. The variation of the cross-pol radiation levels within bandwidth suggests that for an antenna with an achieved broad impedance bandwidth, it is important to check its corresponding radiation performance such as the co-to-cross-pol ratios across the entire bandwidth of interest. For some applications with specific radiation requirements, the degraded radiation performance may offset the advantage of having broad impedance bandwidth, and may even limit the application of the antenna.

1.3.2 BENT PLATE ANTENNAS

Another typical plate design is a bent plate monopole. Investigations have shown the effects of the geometry of the radiator on the impedance and radiation characteristics of the antenna.[13] Figure 1.10 ia a schematic of a typical bent plate monopole. Consider a square radiating PEC sheet of dimensions $l \times l = 70$ mm $\times$ 70 mm. The feed gap g is 5 mm. The radiator consists of the horizontal and vertical portions with lengths l_h and l_v, respectively. A cylindrical 50-Ω probe with a 0.6 mm radius excites the midpoint of the bottom of the sheet through an infinite ground plane.

Figure 1.11 shows the co-to-cross-pol ratios in the x–z plane against the ratios l_v/λ_l of the vertical length to the wavelength at the lower edges of impedance bandwidths. Usually, the minimum size of an antenna is determined by the lower edge frequency where the antenna is

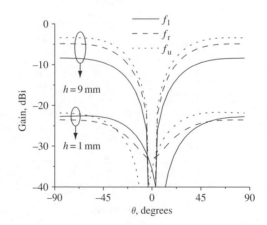

Figure 1.9 Cross-pol radiation patterns in the H-plane at f_l, f_r and f_u for $h = 1$ mm and 9 mm.

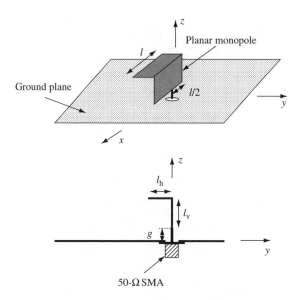

Figure 1.10 A bent plate antenna with size $l \times l = 70\,\mathrm{mm} \times 70\,\mathrm{mm}$ and $l_\mathrm{h} + l_\mathrm{v} = 70\,\mathrm{mm}$.

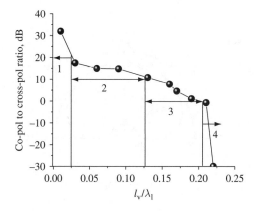

Figure 1.11 Co-to-cross-pol ratios in the $x\text{--}z$ plane versus ratios l_v/λ_1 for the antenna shown in Figure 1.10.

well matched to its feeder. It is evident that the co-to-cross-pol ratios decrease from 33 dB to $-30\,\mathrm{dB}$ as the ratio l_v/λ_1 increases from 0.03 to 0.22. In particular, in terms of the radiation features of the antenna operating at its dominant modes, four categories can be approximately but reasonably constructed:

- microstrip patch antennas with $l_\mathrm{v} < 0.03\lambda_1$ (category 1)
- suspended plate antennas with $0.03\lambda_1 \leq l_\mathrm{v} \leq 0.12\lambda_1$ (category 2)
- planar inverted-L/F antennas with $0.12\lambda_1 \leq l_\mathrm{v} \leq 0.20\lambda_1$ (category 3)
- planar monopole antennas with $l_\mathrm{v} > 0.20\lambda_1$ (category 4).

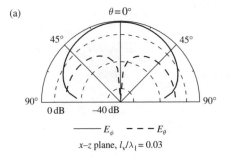

(a)

x–z plane, $l_v/\lambda_1 = 0.03$

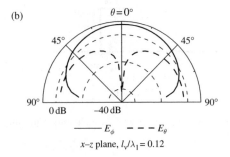

(b)

x–z plane, $l_v/\lambda_1 = 0.12$

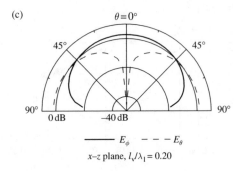

(c)

x–z plane, $l_v/\lambda_1 = 0.20$

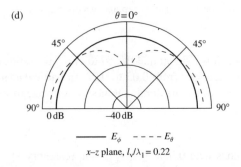

(d)

x–z plane, $l_v/\lambda_1 = 0.22$

Figure 1.12 Radiation patterns for E_θ and E_ϕ components in the x–z plane.

It should be noted that the co-to-cross-pol ratios are also associated with the radiator shape, the aspect ratio, the feed-point location and the substrate supporting the radiating patch. For example, square microstrip antennas with 1:1 aspect ratios usually have small co-to-cross-pol ratios.

Figure 1.12 plots the radiation patterns for E_ϕ (co-pol) and E_θ (cross-pol) components in the x–z plane. The dimensions (l_v/λ_1) of the antennas are chosen such that they fall well into the four categories. The figures clearly show the variation in the radiation properties of the bent planar antennas with varying length l_v/λ_1.

Figure 1.13 shows the input impedances of the antennas analysed in Figure 1.12. The antennas with smaller lengths (l_v/λ_1) have higher input impedances and Q values. As l_v/λ_1 approaches zero, the radiator operates as a microstrip antenna as shown from the radiation patterns in Figure 1.12(a). As l_v/λ_1 becomes large enough (for example, more than 0.20), the radiator operates as a monopole antenna as shown from the radiation patterns in Figure 1.12(d).

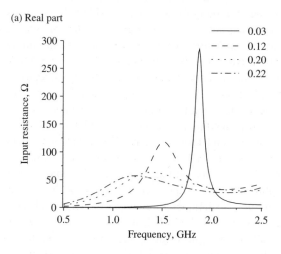

(a) Real part

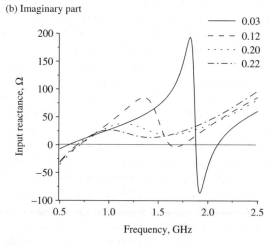

(b) Imaginary part

Figure 1.13 Input impedance versus frequency for $l_v/\lambda_1 = 0.03, 0.12, 0.20$ and 0.22.

It is well known that microstrip patch antennas and monopoles with distinct radiation characteristics are the most basic radiators. The former type operating in its dominant mode radiates the θ components of the electric field mainly at the boresight (z-axis direction), and the latter in the x–y plane shown in Figure 1.10. From Figure 1.12 it is clear that the suspended plate antenna and planar inverted-L or inverted-F antennas are situated between microstrip patch antennas and planar monopole antennas. Thus, for suspended plate antennas having a broad impedance bandwidth, efforts should be made to enhance their radiation performance when used as broadband microstrip patch antennas.

1.4 OVERVIEW OF THIS BOOK

Based on the discussion above, planar antennas have been categorized into microstrip patch antennas (MPAs), suspended plate antennas (SPAs), planar inverted-F/L antennas (PIFAs/PILAs) and planar monopole antennas (PMAs). Techniques for enhancing the bandwidths of various planar antennas are elaborated in the following chapters.

In Chapter 2, the important features of microstrip antennas are introduced by examining a typical rectangular MPA. After that, broadbanding techniques for MPAs with thin dielectric substrates are reviewed. They mainly include lowering the Q value, using matching networks, and introducing multiple resonances. Two design examples with an impedance-matching stub and stacked elements are given to demonstrate the techniques.

Chapter 3 discusses suspended plate antennas with thick dielectric substrates of very low permittivity, or without any dielectric substrate. These antennas feature broad impedance bandwidths of around 8 % for VSWR = 2. Techniques are described to further enhance the impedance bandwidth, such as the use of a capacitive load, slotting the radiators, and electromagnetic coupling. Many examples are used to show the procedures.

Chapter 3 also introduces techniques to alleviate the degraded radiation performance of suspended plate antennas. The method of dual probes is described first, with an example. Then, a probe-fed half-wavelength feeding structure is used to enhance the radiation performance. After that, a center probe-fed slot feeding structure is shown to feature satisfactory radiation performance within a broad bandwidth. A broadband suspended plate antenna with double L-shaped probes is used as the case study. Finally, shorting strips and slots applied in suspended plate antennas are investigated.

Chapter 3 also discusses the mutual coupling between suspended plate antennas, and arrays with the suspended plate elements on tiered ground planes.

In Chapter 4, planar inverted-F/L antennas with broad impedance bandwidths are reviewed and some new techniques introduced, with case studies. The applications of planar inverted-F antennas in handphones and laptop computers is examined. In particular, the text addresses planar inverted-F antennas with a small system ground plane in handphone design.

Chapter 5 introduces planar monopoles and their applications. A planar monopole with the radiator standing above the ground plane or printed on a board (PCB) may be the simplest broadband design. Its low fabrication cost and broad impedance and radiation performance are very attractive to antenna engineers. In particular, the emerging radar systems based on ultra-wideband technology are boosting the development of planar monopoles.

Finally, issues related to applications of planar monopoles (including Vivaldi antennas, a type of planar horn antenna), are addressed with two case studies.

REFERENCES

[1] R. F. Harrington, *Time-harmonic Electromagnetic Fields*. New York: McGraw Hill, 1961.

[2] R. E. Collin and S. Rothschild, 'Evaluation of antenna Q,' *IEEE Transactions on Antennas and Propagation*, vol. 29, no. 1, pp. 23–27, 1964.

[3] R. L. Fante, 'Quality factor of general ideal antenna,' *IEEE Transactions on Antennas and Propagation*, vol. 17, no. 2, pp. 151–155, 1969.

[4] D. M. Grimes and C. A. Grimes, 'Bandwidth and Q of antennas radiating TE and TM modes,' *IEEE Transactions on Antennas and Propagation*, vol. 43, no. 2, pp. 217–226, 1995.

[5] A. C. Ludwig, 'The definition of cross polarization,' *IEEE Transactions on Antennas and Propagation*, vol. 21, no. 1, pp. 116–119, 1973.

[6] P. S. Hall, 'Probe compensation in thick microstrip patches,' *Electronics Letters*, vol. 23, pp. 606–607, 1987.

[7] K. Levis, A. Ittipiboon and A. Petosa, 'Probe radiation cancellation in wideband probe-fed microstrip arrays,' *Electronics Letters*, vol. 36, no. 7, pp. 606–607, 2000.

[8] A. Petosa, A. Ittipiboon and N. Gagnon, 'Suppression of unwanted probe radiation in wideband probe-fed microstrip patches,' *Electronics Letters*, vol. 35, pp. 355–357, 1999.

[9] Z. N. Chen and M. Y. W. Chia, 'Enhanced radiation performance of a suspended plate antenna,' *Proceedings of International Conference on Electromagnetic and Advanced Applications (ICEAA), Torino, Italy*, vol. 1, September 2001.

[10] Z. N. Chen and M. Y. W. Chia, 'Center-fed microstrip patch antenna,' *IEEE Transactions on Antennas and Propagation*, vol. 51, no. 3, pp. 483–487, 2003.

[11] Z. N. Chen and M. Y. W. Chia, 'A novel center-fed suspended plate antenna,' *IEEE Transactions on Antennas and Propagation*, vol. 51, no. 6, pp. 1407–1410, 2003.

[12] Z. N. Chen and M. Y. W. Chia, 'Experimental study on radiation performance of probe-fed suspended plate antenna,' *IEEE Transactions on Antennas and Propagation*, vol. 51, no. 8, pp. 1964–1971, 2003.

[13] Z. N. Chen, 'Note on impedance characteristics of L-shaped wire monopole antenna,' *Microwave and Optical Technology Letters*, vol. 26, no. 1, pp. 22–23, 2000.

2

Broadband Microstrip Patch Antennas

2.1 INTRODUCTION

Microstrip antennas and arrays have attracted much attention from researchers and engineers and have been used extensively in RF and microwave systems, such as communications, radar, navigation, remote sensing, and biomedical systems. Microstrip antennas can take a variety of forms, such as patch, dipole, slot, or a traveling-wave structure, designed for specific applications. Examples are illustrated in Figures 2.1–2.3.

Patch or patch-like radiators feature unique characteristics and have been studied in depth. They offer several important advantages:

- *Low profile*. The thickness of a microstrip patch antenna is usually less than $0.03\,\lambda_o$ (λ_o is the operating wavelength in free space).
- *Light weight*. A microstrip patch antenna is usually made from perfectly electrically conducting (PEC) foil affixed on a dielectric substrate.
- *Conformability to surfaces of substrates*. A microstrip patch antenna may be of a planar or nonplanar surface, which can completely conform to the surface of the dielectric substrate it is attached to.
- *Low cost*. A microstrip patch antenna is commonly fabricated using an inexpensive printed-circuit technique. The substrate is usually the most costly portion of the antenna.
- *Integration with other circuits*. It is easy to completely integrate a microstrip patch antenna on a printed-circuit board (PCB) with other planar circuits.
- *Versatility*. A microstrip patch antenna is very versatile in terms of impedance, resonant frequency, radiation pattern, polarization and operating mode, by choice of shape and feeding arrangement. Many techniques, such as adding shorting pins, varactor diodes, loading and slotting the patch, or introducing parasitic elements, can be applied to the antenna.

Broadband Planar Antennas: Design and Applications   Zhi Ning Chen and Michael Y. W. Chia
© 2006 John Wiley & Sons, Ltd

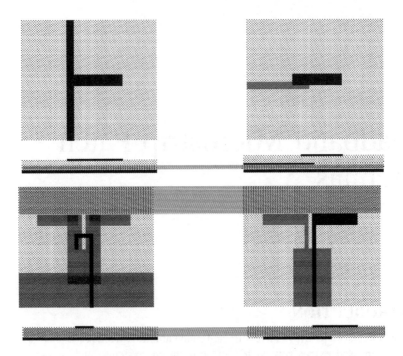

Figure 2.1 Printed dipole antennas.

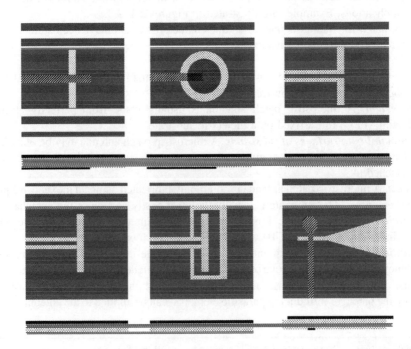

Figure 2.2 Printed slot antennas.

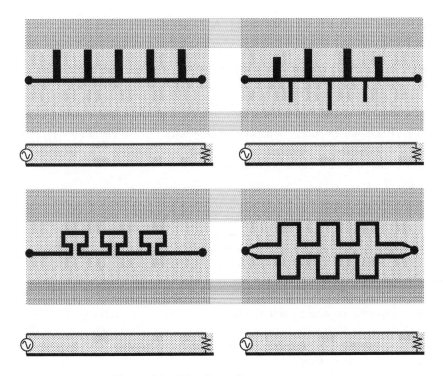

Figure 2.3 Printed traveling-wave antennas.

However, microstrip patch antennas in their basic forms also suffer from some drawbacks, such as narrow impedance bandwidth (typically of around 1%), poor polarization purity, low radiation efficiency, poor power capability, poor scan performance, and excitation of surface waves. The narrow impedance and radiation bandwidths are the primary disadvantages and, are caused mainly by frequency-dependent efficiency, radiation direction or polarization; this essentially limits the application of this type of antenna. For this reason, much effort has been devoted to the development of broadband techniques, as emphasized in this chapter.

Microstrip patch antennas are one of the most basic and important types of planar antenna. Many of the concepts and techniques used with microstrip patch antennas can be applied directly to other planar antennas. They have been widely described in books[1–9] and review papers in international journals.[10–15] If readers are interested in the fundamentals, it would be useful to study these books and papers in depth.

This chapter reviews the techniques for improving the bandwidth of a microstrip patch antenna. First, the basic features of these antennas are reviewed briefly. A rectangular patch antenna is chosen to show the basic impedance and radiation characteristics, after which broadband techniques are described concisely. The techniques can be applied also to other planar antennas, either directly or after some modifications. Lastly, a practical design procedure for a broadband dual-polarization antenna is given as an example.

2.2 IMPORTANT FEATURES OF MICROSTRIP PATCH ANTENNAS

The basic configuration is a metallic patch etched on to an electrically thin and grounded dielectric substrate, as shown in Figure 2.4. The thickness of the substrate is much less than λ_o. In general, the patch element is fed asymmetrically by an unbalanced feed. The main dimensions of patches are of the order of $\lambda_e/2$, where λ_e is an equivalent operating wavelength. The radiation pattern is dependent on electric current distributions on the patch. The patch shape, feeding structure and substrate properties can be chosen to achieve the desired performance for a specific application.

2.2.1 PATCH SHAPES

The radiator should be a material with low ohmic loss and high conductivity at the operating frequency (such as copper), which can be fixed to a dielectric substrate. The shape can be an ordinary rectangle, square, ellipse, circle, triangle, ring, pentagon, or their variations; Figure 2.5 shows some typical radiators. More complex variations on the basic shapes are frequently used to meet particular design demands. The selection of a particular shape is contingent on specific requirements in terms of polarization, bandwidth, gain, etc. Generally, the antenna's characteristics are defined by the excited operating modes, which depend on the shape and dimensions of the patch, the thickness and dielectric constant of the substrate, as well as the feed arrangement.

2.2.2 SUBSTRATES

There is a multitude of dielectric materials available for substrates. Important parameters are the dielectric constant ($2.2 \le \epsilon_r \le 16$ in RF or microwave bands), the dielectric loss tangent ($0.0001 \le \tan \delta \le 0.06$) or imaginary part of the dielectric constant, and cost. Due to their low cost, ease of manufacture and good surface adhesion, plastics are commonly used in RF and

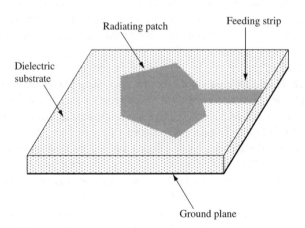

Figure 2.4 A microstrip patch antenna with a radiator of an arbitrary shape.

Figure 2.5 Shapes of radiators.

microwave bands, although they have large thermal expansion coefficients, poor dielectric properties, poor dimensional stability and poor thermal conductivity compared with other materials such as ceramic and sapphire.

Another consideration when choosing the substrate material is the effect of the dielectric constant on the radiation characteristics. A high dielectric constant usually results in low radiation from a microstrip patch antenna.

2.2.3 FEEDING STRUCTURES

The feed must transfer RF or microwave energy efficiently from the transmission system to the antenna. The design of the feeding structure directly governs the impedance matching, operating modes, spurious radiation, surface waves and geometry of the antenna or array. The feeding structure thus plays a vital role in widening the impedance bandwidth and enhancing radiation performance. In this sense, development of broadband techniques is highly reliant on the improvement of feeding structures, and this has been reflected by the progress in the design of other types of antenna besides microstrip patch antennas.

Figure 2.6(a) shows a feeding structure comprising a probe, which extends from a coaxial line and through a ground plane. Such a coaxial probe is commonly used to excite microstrip patch antennas. Usually, a coaxial connector is soldered to the ground plane below a dielectric substrate. By adjusting the location of the feed point, impedance matching between the coaxial line and the radiating patch will be achieved. However, this feeding structure is *not* suitable for arrays owing to the great number of solder points and coaxial lines. Other limitations include the increase in spurious radiation, surface waves and input inductance for antennas with thick substrates.

To reduce the cost and simplify the design, the feeding structure, microstrip patch antenna and even other circuits can be fabricated on the same surface, as illustrated in Figure 2.6(b). The feeding strip may be fed by a transmission line, a surface-mounted RF connector or a

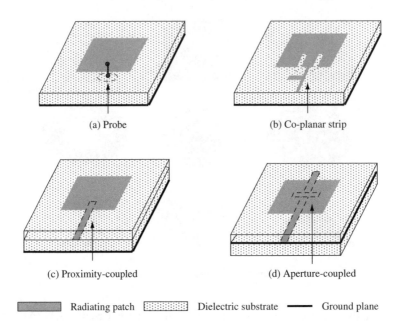

(a) Probe   (b) Co-planar strip

(c) Proximity-coupled   (d) Aperture-coupled

▬▬▬ Radiating patch   ▦▦▦ Dielectric substrate   ▬▬ Ground plane

Figure 2.6 Types of feeding structure used in microstrip patch antennas.

coaxial line at the edge of the ground plane. The feeding strip may be directly attached or coupled to the edge of the patch. This configuration is conducive to array applications. Many planar impedance-matching techniques – such as notching the patch near the feeding strip or adding an impedance-matching network (e.g. stubs) between the strip and the patch – can be used to achieve good matching between the feed and the patch.

Figure 2.6(c) depicts a proximity-coupled feeding structure, whereby the feeding strip is located below the radiating patch but above a ground plane. There are thus two dielectric layers between the patch and ground plane. Energy is transferred by means of the electromagnetic coupling between the patch and the feeding strip. Owing to the double-layered structure, the impedance bandwidth may be increased by properly aligning the patch and the feeding strip. However, the multi-layered structure increases fabrication cost and surface wave losses.

An aperture-coupled structure is an important feeding configuration in microstrip patch antennas, as shown in Figure 2.6(d). Two substrates are separated by a common ground plane; a radiating patch is located on the upper substrate while a feeding strip is on the lower substrate. A nonresonant aperture, usually a narrow slot situated between the patch and the feeding strip, is cut from the ground plane. The feeding strip is electromagnetically coupled to the patch through the aperture. The coupling between the patch and the feeding strip is controlled by the aperture size and shape. Impedance matching can be achieved by optimizing the aperture parameters, the location and length of the strip, and the dielectric constants of the two substrates. An antenna excited by this feeding structure usually features a broad impedance bandwidth and high polarization purity, although it has higher fabrication complexity and cost than single-layered structures.

Another feeding scheme is based on the co-planar waveguide (CPW). This scheme is very suited to microstrip patch antenna design, and has been widely used, for example, in

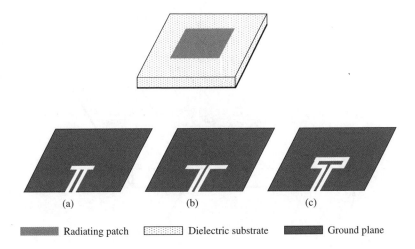

(a)   (b)   (c)

Radiating patch   Dielectric substrate   Ground plane

Figure 2.7   Co-planar wageguide (CPW) feeding structures.

microwave monolithic integrated circuits (MMICs). The waveguide is cut on the ground plane, and slots with various shapes etched at the end of the CPW are employed to achieve impedance matching, as shown in Figure 2.7. The coupling between the CPW and patch is capacitive for Figure 2.7(a) and inductive for Figure 2.7(b). However, as with aperture coupling, the slots on the ground plane cause high back radiation; typically, the front-to-back ratio is about 10 dB. To suppress the unwanted back radiation, a loop of any shape is introduced as illustrated in Figure 2.7(c).

In CPW structures, the odd mode excited in the coupled slot line is dominant. In this odd mode, the equivalent magnetic currents in adjacent CPW slots are out of phase. Therefore, the radiation from the feeding structure is negligible. This feature makes it suitable for array applications, where the back radiation from, as well as the mutual coupling between, the feeding structures are alleviated greatly.

2.2.4 EXAMPLE: RECTANGULAR MICROSTRIP PATCH ANTENNAS

The rectangular shape is widely used in practice because of its simple shape and attractive characteristics. There are various analytical approaches to modeling a rectangular microstrip patch antenna. Both the transmission line model (TLM) and cavity model (CM) are very popular and practical.[16-24] Compared with numerical methods, the analytical methods usually provide more usable physical patterns.

In the TLM, the antenna is considered fundamentally as an open-ended section of transmission line of length L and width W, as shown in Figure 2.8. The terms t and ϵ_r denote the thickness and relative dielectric constant of the substrate, respectively. The resonant frequency for its transverse magnetic (TM_{m0}) mode can be evaluated from

$$f_{rm} = \frac{mc}{2(L + \Delta L)\sqrt{\epsilon_{reff}}} \qquad (2.1)$$

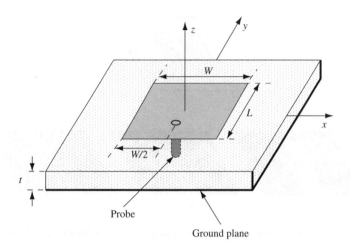

Figure 2.8 A rectangular microstrip patch antenna.

where c is the velocity of light (2.998×10^8 m/s); m is an integer ($\neq 0$); L is the length of the patch; and ΔL is the equivalent length after taking account of the fringing fields at the two open ends. The dominant mode is TM_{10}.

$$\frac{\Delta L}{t} = 0.412 \frac{(\epsilon_{reff} + 0.3)(W/t + 0.264)}{(\epsilon_{reff} - 0.258)(W/t + 0.8)}. \tag{2.2}$$

Here, ϵ_{reff} is the *effective* relative dielectric constant and is given by

$$\epsilon_{reff} = \frac{\epsilon + 1}{2} + \frac{\epsilon - 1}{2}\left(1 + 10\frac{t}{W}\right)^{-\alpha\beta} \tag{2.3}$$

where

$$\alpha = 1 + \frac{1}{49}\log\frac{\left(\frac{W}{t}\right)^4 + \left(\frac{1}{52}\frac{W}{t}\right)^2}{\left(\frac{W}{t}\right)^4 + 0.432} + \frac{1}{18.7}\log\left[1 + \left(\frac{1}{18.1}\frac{W}{t}\right)^3\right]$$

$$\beta = 0.564\left(\frac{\epsilon_{reff} - 0.9}{\epsilon_{reff} + 3}\right)^{0.053}.$$

In the CM, a microstrip antenna with a metallic patch and a ground plane can be approximated by a dielectric-filled cavity. In principle, the CM can handle any microstrip antennas with arbitrary shapes. However, owing to the mathematical complexity, it is used to analyse microstrip patch antennas with regular shapes, such as rectangles, ellipses, circles, or triangles. The effects of the higher-order modes can be taken into account. The resonant frequencies of the possibly excited higher-order modes TM_{mn} can be determined from

$$f_{rmn} = \frac{c}{2\sqrt{\epsilon_{reff}(L, W)}}\sqrt{\left\{\left[\frac{m}{L + 2\Delta L(W)}\right]^2 + \left[\frac{n}{W + 2\Delta W(L)}\right]^2\right\}} \tag{2.4}$$

where f_{rmn} is the resonant frequency for the TM_{mn} mode.[24]

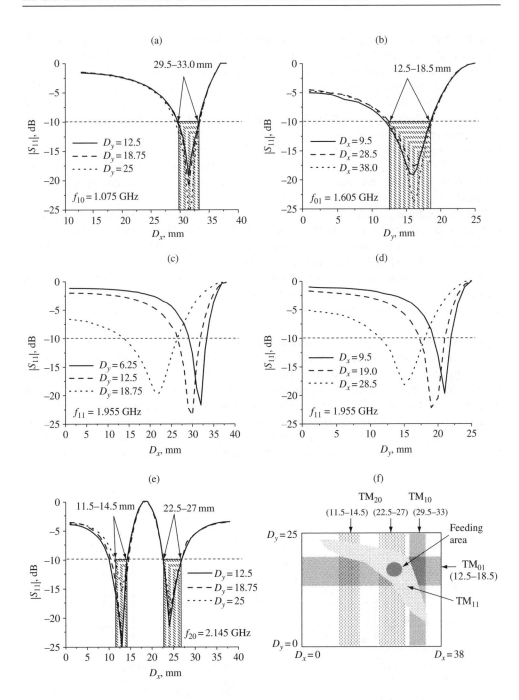

Figure 2.9 The well-matched region of a rectangular microstrip patch antenna (dimensions are in millimeters).

In the probe-fed rectangular microstrip antenna shown in Figure 2.8, the patch width W is 76 mm and the length L is 50 mm. The thickness t of the dielectric substrate is 60 mil, and the relative dielectric constant ϵ_r is 3.38. The patch is fed by a 1.2-mm-thick cylindrical probe at position (D_y, D_x) as shown in Figure 2.9(f). The first four lower modes, namely TM_{10}, TM_{01}, TM_{11} and TM_{20}, have been predicted by equation 2.4; they are within the frequency range from 0.5 GHz to 2.5 GHz. By properly selecting the location of the feed point, the modes can be excited simultaneously. Figures 2.9(a)–(e) provide the matching conditions for each mode according to the location of the feed point. Figure 2.9(f) illustrates the well-matched regions for each mode. With a tradeoff between the excitation of the modes, the optimum feed region is obtained.

With the feed point located at $D_x = 29$ mm and $D_y = 15$ mm, the input impedance and return loss $|S_{11}|$ for frequencies ranging from 0.5 GHz to 2.5 GHz are shown in Figure 2.10. Within the frequency range, the first four modes – namely TM_{10} (1.075 GHz), TM_{01} (1.605 GHz), TM_{11} (1.955 GHz) and TM_{20} (2.145 GHz) – are well excited.

The surface electric currents and radiation patterns at the first four resonant frequencies are plotted in Figures 2.11–2.14. In the figures, the maximum gain is achieved in the $\phi = 0°$ and 90° cuts. It should be noted that the electric fields are normalized by the maxima in each of the cuts.

By comparing the imaginary (quadrature) and real (in-phase) components of the electric currents on the patches at the resonant frequencies, it is seen that the quadrature components are hardly dependent on the location of the feed point at the resonances, whereas the position of the feed point affects the in-phase components to a great extent. Moreover, the magnitudes of the imaginary components are several times greater than those of the real components. In other words, the imaginary components contribute primarily to the radiation at the resonances.[25] Therefore, in the following study, only the imaginary components of the currents induced on the radiators will be considered in the discussions on radiation characteristics.

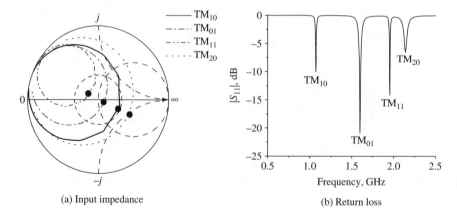

(a) Input impedance   (b) Return loss

Figure 2.10   The input impedance and return loss $|S_{11}|$.

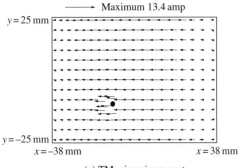

(a) TM_{10} imaginary part

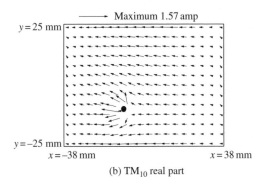

(b) TM_{10} real part

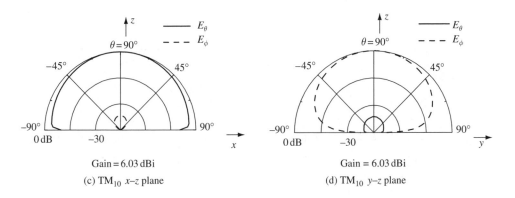

Gain = 6.03 dBi   Gain = 6.03 dBi

(c) TM_{10} x–z plane   (d) TM_{10} y–z plane

Figure 2.11 Current distributions and radiation patterns for the TM_{10} mode.

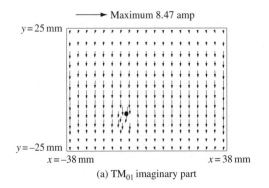

(a) TM$_{01}$ imaginary part

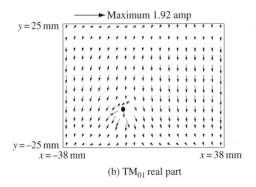

(b) TM$_{01}$ real part

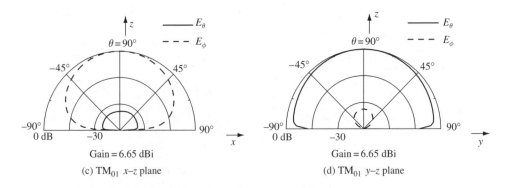

(c) TM$_{01}$ x–z plane   (d) TM$_{01}$ y–z plane

Figure 2.12   Current distributions and radiation patterns for the TM$_{01}$ mode.

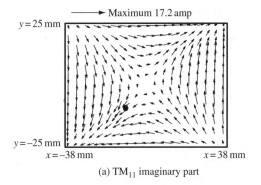

(a) TM_{11} imaginary part

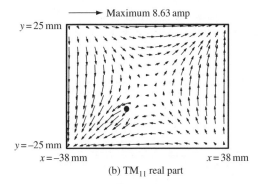

(b) TM_{11} real part

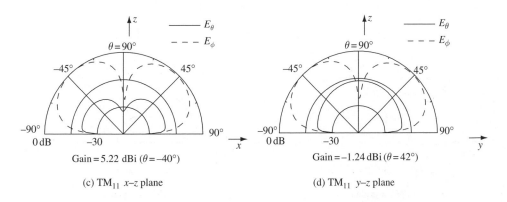

(c) TM_{11} x–z plane

(d) TM_{11} y–z plane

Figure 2.13 Current distributions and radiation patterns for the TM_{11} mode.

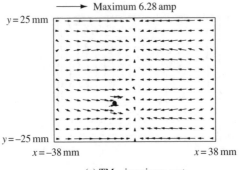

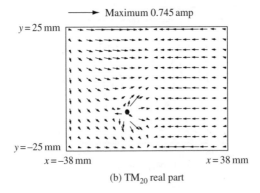

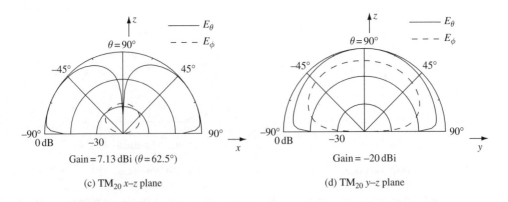

Figure 2.14 Current distributions and radiation patterns for the TM$_{20}$ mode.

2.3 BROADBAND TECHNIQUES

A variety of broadband techniques have been developed using the three approaches categorized in Table 2.1. It is known that the factors affecting the bandwidth of a microstrip patch antenna are primarily the shape of the radiator, the feeding scheme, the substrate and the arrangements of radiating and parasitic elements. Essentially, the broad bandwidth of a microstrip patch antenna can be attributed to its low Q value and simultaneously well-excited multiple resonances. If the antenna is considered as a high-Q filter, lowering the Q by reducing the energy around the radiator or increasing losses broadens the bandwidth at its resonance. Alternatively, by inserting a broadband impedance network between the antenna and the feeder, good matching over a broad frequency range can be attained. If two or more adjacent modes are well excited simultaneously, the bandwidth can be twice or more than that for the single resonance.

2.3.1 LOWERING THE Q

As mentioned in Chapter 1, a microstrip patch antenna having a larger sphere suffers from a narrow bandwidth as the whole volume of the enclosing sphere is not utilized effectively. Therefore, a microstrip patch antenna can be considered as a high-Q circuit, so one way to alleviate the narrow bandwidth problem is to reduce the Q.

Investigations have shown that the shape of a radiator affects the impedance bandwidth, even for the same maximum dimensions. However, the improvement in the bandwidth is quite limited.[7,9] For example, the bandwidths for return loss $|S_{11}|$ less than -10 dB of the antenna shown in Figure 2.15 with different aspect ratios (W/L) are listed in Table 2.2. The thickness and dielectric constant of the substrate are $t = 1.52$ mm and $\epsilon_r = 3.38$. The length L is fixed at 54 mm while the width W varies from 10 mm to 108 mm. The resonant frequency is around 1.5 GHz. It is clear that the bandwidth is affected by the geometry of the antenna due to the lower Q from the larger size. The impedance bandwidth for $|S_{11}| < -10$ dB is still around 1 %.

The shape of a radiator, which affects the operating modes is critical for its radiation performance. Therefore, this technique is hardly employed in practical designs.[26]

Table 2.1 Broadband techniques for microstrip patch antennas.

Approach	Techniques
Lower the Q	Select the radiator shape Thicken the substrate Lower the dielectric constant Increase the losses
Use impedance matching	Insert a matching network Add tuning elements Use slotting and notching patches
Introduce multiple resonances	Use parasitic (stacked or co-planar) elements Use slotting patches, insert impedance networks Use an aperture, proximity coupling

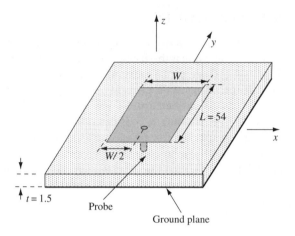

Figure 2.15   A probe-fed rectangular microstrip patch antenna (dimensions in millimeters).

Table 2.2   Bandwidth against aspect ratio for the antenna in Figure 2.15.

W/L	10/54	27/54	54/54	81/54	108/54
Bandwidth for $|S_{11}| < -10\,\text{dB}$	0.38 %	0.68 %	0.70 %	0.95 %	1.2 %

As an alternative, a thick substrate with low dielectric constant is good for improving the impedance bandwidth of these antennas.[7,27–34] This monotonically reduces the Q and broadens the bandwidth.

Another important consideration is the increase in losses due to undesired surface waves, which lowers the radiation efficiency, excites spurious radiations, and degrades the radiation patterns. Investigations have shown that the impedance bandwidth monotonically increases with the substrate thickness while radiation efficiency rapidly decreases.[7] However, the achievable bandwidth will decrease when the thickness exceeds a certain value. For example, the achieved bandwidth shown in Figure 2.16 will drop when the thickness ratio t/λ_0 is greater than about 0.058. In a probe-fed microstrip patch antenna, the long probe leads to high input inductance and resistance around its resonant frequency. The increased input impedance leads to difficulty in impedance matching. Figure 2.16 shows that a microstrip patch antenna supported by a substrate with a lower dielectric constant of $\epsilon_r = 3.38$ has a broader bandwidth than that with $\epsilon_r = 10.2$, for the same thickness ratio t/λ_0.

2.3.2 USING AN IMPEDANCE MATCHING NETWORK

An impedance matching network can be introduced to realize good matching between a radiator having frequency-dependent impedance and a feed structure with a constant characteristic impedance. This naturally leads to a broader bandwidth. There are two common methods employed in microstrip patch antennas. One is to insert a separate matching network without altering the radiator. Another is to introduce an on-patch matching network either

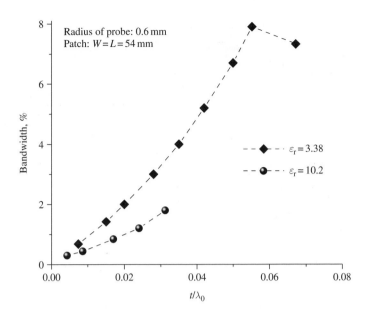

Figure 2.16   Bandwidth versus substrate thickness ratio.

by slotting or notching the radiator. With both methods, the insertion of a lossy or lossless impedance matching network between the antenna and feeding structure can directly improve the impedance bandwidth. This method has been commonly used in a variety of antenna applications.[35–37]

The impedance network can be applied in various ways, as shown in Figure 2.17. For the planar feeding structures in (a) and (b), the matching network and the feeding strip can be etched on the same plane. The matching network can be a quarter-wavelength impedance transformer, tuning stubs, active components or their variations, as shown in Figure 2.18. The advantage of this is that the radiator need not to be changed, which simplifies the design by allowing the impedance matching and radiation performance to be controlled independently. The main drawback is the increase in size, which make this method difficult to be applied in arrays. Also, the additional circuit reduces the system's efficiency due to increased losses from spurious radiation.

The off-board network arrangement shown in Figure 2.17(c) suppresses the effect of spurious radiation on the radiation performance without any increase in size. However, for a probe-fed microstrip patch antenna supported by a thick dielectric substrate, the input inductance is high. Figure 2.19 shows some capacitive circuits inserted between the probe and the patch to compensate for the extra inductance.[38–42] In (a)–(c), the strong coupling between the patch and the circular top-hat at the end of the probe produces the capacitance refined to tune the input impedance. The circular disc in (d) forms a frill of the probe.

Insertion of an on-patch matching network consisting of a slot or notch on the radiating patch forms an inset on the radiating patch to transfer the higher impedance at the radiating edge of the patch to the lower 50-Ω characteristic impedance of the feeding structure.[43–46] Owing to the simple design and no increase in size, such inset-fed microstrip patches are used

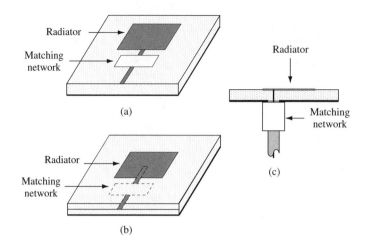

Figure 2.17 Theoretical impedance matching networks.

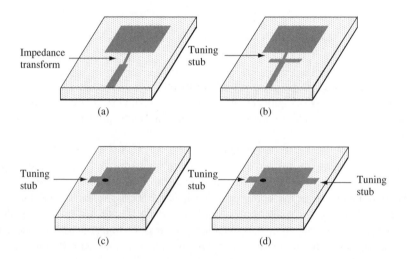

Figure 2.18 Practical impedance matching networks for planar structures.

extensively, particularly in arrays and MMIC designs. Figure 2.20 shows several inset-fed patches for broadband impedance matching, where feeding structures may be striplines or probes.

2.3.3 CASE STUDY: MICROSTRIP PATCH ANTENNA WITH IMPEDANCE MATCHING STUB

Consider a square patch antenna of dimensions 54 mm × 54 mm, $t = 15$ mm, and $\epsilon_r = 3.38$, which has a well-matched bandwidth (for $|S_{11}| < -10$) of 7.3 % (1.32–1.42 GHz) and the

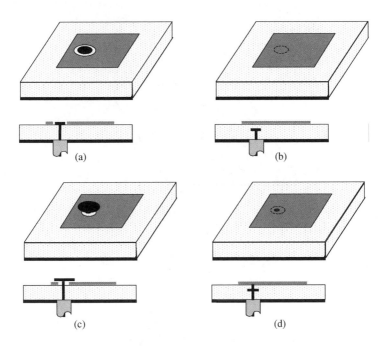

Figure 2.19 Impedance matching networks for probe-fed structures.

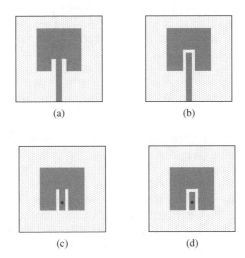

Figure 2.20 Impedance matching networks with slots and notches.

same geometry as shown in Figure 2.15. Owing to the high input impedance, the impedance matching is degraded as illustrated in Figure 1.6, and it is impossible to improve the impedance matching by shifting the feed point to the edge because the feed point is already located at the radiating edge. In this design example, a tuning stub as shown in Figure 2.18(c) is employed for broadband impedance matching.

(a) Geometry

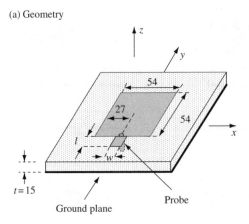

(b) Input impedance loci for varying stub lengths

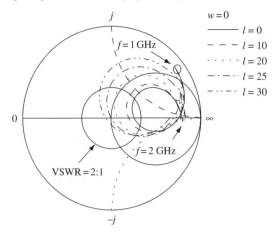

Figure 2.21 The square patch antenna with a stub (dimensions in millimeters).

The feed point is fixed at the midpoint of the radiating edge. A rectangular stub of width $w = 8$ mm and length l is centrally connected to the edge, and close to the feed point as depicted in Figure 2.21(a). Figure 2.21(b) shows the input impedance loci on a Smith chart for various stub lengths. Increasing the length of the stub from 0 mm to 30 mm greatly enlarges the locus loop with a reduction in the input impedance. The impedance matching is optimal when the stub length is 25 mm. The achieved bandwidths for $|S_{11}|$ less than -10 dB are given in Table 2.3. The bandwidth increases by 18 % with a decrease in the resonant frequency from 1.37 GHz to 1.16 GHz.

It should be noted that, unlike the antenna fed by a co-planar stripline, where a quarter-wavelength impedance transformer can be used to achieve impedance matching, the width of the additional stub used in the probe-fed antenna does not need to meet the requirement for the quarter-wavelength impedance transformer. This example also reveals that the selection of the stub width w is more flexible.

Table 2.3 Bandwidths for varying stub lengths *l*.

l mm	0	10	20	25	30
Bandwidth for $|S_{11}| < -10\,$dB (%)	7.3	7.6	8.1	8.6	7.4
Frequency range (GHz)	1.32–1.42	1.27–1.37	1.18–1.28	1.11–1.21	1.04–1.12

2.3.4 INTRODUCING MULTIPLE RESONANCES

Microstrip patch antennas are basically structures operating at the resonant frequncies with a very limited bandwidth. Operation at adjacent multiple resonances has been shown to be a practical way of enhancing the bandwith. Using this approach, two or more adjacent resonances are well excited within the operating frequency range simultaneously. This technique has been commonly used in RF circuits such as stage-tuned filters and other antenna applications, such as log-periodic toothed planar antennas and log-periodic dipole arrays.

The most direct way is to introduce additional radiating patches located close to the main radiator. These parasitic patches are excited by means of the electromagnetic coupling between them and the main patch. The parasitic patches can be in the same plane as, or stacked above, the main patch.

Co-planar Arrangements

Figure 2.22 shows co-planar arrangements, whereby one or more parasitic elements are located around the driven element. The small gaps allow the parasitic elements to be strongly coupled to the driven element. The elements have almost the same size. The well-excited resonances are adjacent to one another such that a broader well-matched bandwidth is achieved, which is usually a few times broader than that of the single radiating element.

In arrangements (a) and (b), the parasitic elements are coupled to the main patches along the non-radiating edges.[47–50] Choice of the feeding structure is flexible, as probes, striplines and CPWs are all applicable. One drawback is the increased cross-polarized radiation levels

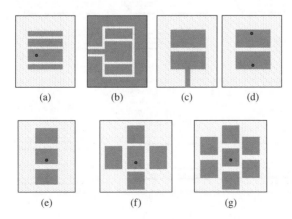

Figure 2.22 Co-planar coupling schemes.

in H planes, and even co-polarized radiation patterns in E planes, especially at higher operating frequencies.

In arrangements (c)–(e), the parasitic elements are coupled to the main patch along the main radiating edges.[51-53] The improved impedance bandwidth can be five times broader than that of the single radiating element. Such configurations also feature unchanged radiation performance in E planes, but distorted co-polarized radiation patterns in H planes at higher operating frequencies.

The arrangement (f) combines the two schemes described above for further bandwidth improvement.[54,55] This scheme can almost double the bandwidth achieved by the other schemes. In addition, this design can be employed to achieve circularly polarized operation by properly positioning the elements around the main patch, as shown in Figure 2.22(g).[55] The main shortcoming of such an arrangement is the significant increase in the lateral size.

Stacked Arrangements

Figure 2.23 shows arrangements whereby parasitic elements are placed above the main patch(es). By dint of the coupling between stacked elements and the driven element, the impedance bandwidth can be increased greatly. Usually, for VSWR $= 1.5$ the bandwidth can reach 10–20 %, particularly as the medium between the upper and bottom patches is air or a material with low permittivity.

Arrangements (a)–(c) are basic stacked microstrip patch antennas.[56-58] The bottom patch is fed directly by either probe or stripline and is usually smaller than the top parasitic element. Typically the bandwidth can be in the order of 10–20 %. In order to increase the gain of the antenna, more than two top elements are stacked right above the bottom one (or with a slight offset) as illustrated in arrangement (d).[59,60] For example, the gain reaches

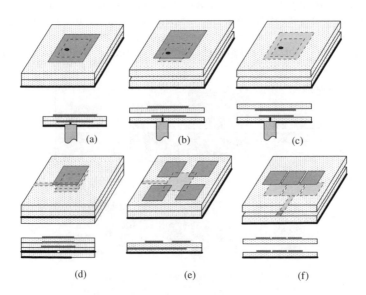

Figure 2.23 Stacked coupling schemes.

8–11.7 dBi for linear polarization and 8–10.6 dBi for circular polarization when two parasitic top elements are placed right above the bottom one.[61] Figure 2.23(d) also features an aperture-coupled feeding structure, whereby the bottom patch is coupled to a stripline under a ground plane through a non-resonant aperture.[59–62] This feeding scheme features broad bandwidth of $>10\%$ and good radiation performance owing to its symmetrical arrangement, but there is undesirable back radiation.

Other possible arrangements include combining the co-planar and stacked structures for a low-profile design with a broad bandwidth and high gain.[63–66] Examples are shown in Figure 2.23(e) and (f). One drawback is the increased lateral size, making design difficult for array applications. However, stacked microstrip antennas and their variations have been applied widely in practical systems, particularly in arrays, due to their broad impedance, good radiation performance and high gain.[67,68]

2.3.5 CASE STUDY: MICROSTRIP PATCH ANTENNA WITH STACKED ELEMENTS

In this design example, we shall consider an antenna for use in an indoor 2.45-GHz WLAN base station. A broadband antenna with 45°-linear polarization is required for various environment where the antenna is to be installed, although the frequency range of the system is from 2.4 GHz to 2.4835 GHz (IEEE 802.116). One of the ways to achieve 45°-linear polarization is to excite one of the corners of the bottom patch. To broaden the impedance bandwidth, four parasitic elements are used and stacked on top of the driven element, as in Figure 2.24. The antenna can be optimized by adjusting the size of the top patches, the spacing between the two dielectric substrates, and the separation of the patches on the top dielectric layer.

The top elements are electromagnetically coupled to a driven patch (33 mm×33 mm) etched on the bottom dielectric layer. The bottom patch is diagonally fed by a 50-Ω microstrip line of width 3.45 mm. The spacing between the two dielectric layers is h. The dielectric

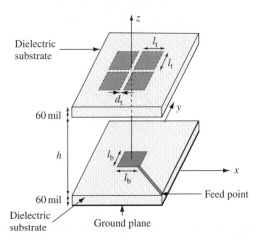

Figure 2.24 Stacked microstrip antenna with four parasitic elements. (Reproduced by permission of John Wiley & Sons, Inc.[69])

substrate (150 mm×150 mm) used for both layers is Roger4003, characterized by a relative permittivity of $\epsilon_r = 3.38$ and thickness 60 mil (about 1.52 mm). The four top patches, each measuring 33 mm×33 mm, are separated by the same gap d_t in both x and y directions.

By computer simulation, the effects of varying the physical parameters on the impedance matching characteristics and radiation performance of the antenna can be examined. The parameters of interests include the patch sizes, the separation between the patches on the top layer, and the spacing between the two layers.

First, the size of the top patches l_t is varied from 32 mm to 35 mm with $l_b = 33$ mm and $h = 7$ mm. According to Figure 2.25, increasing l_t results in excitation of the upper resonance at frequencies higher than 2.6 GHz. Further increases in l_t cause the disappearance of the lower resonance at around 2.4 GHz. Two resonances in terms of VSWR = 1.5 are well excited when $l_t = 33$ mm. By adjusting the size of the top patches, a broad bandwidth can be achieved when both resonances are well-matched and located close to each other. When the minor loops enclose the center of the Smith chart, optimal impedance matching is attained with $l_t = 33$ mm and an impedance bandwidth of 16 % is achieved.

Next, the spacing h between the top and bottom patches is varied with $l_b = l_t = 33$ mm. When h is around 7 mm, two resonances are well-matched as shown in Figure 2.26. A well-matched bandwidth covers a broader frequency range when the two resonances are brought closer together. Further increases in the spacing h lead to a decline in the bandwidth as the lower resonance disappears.

Finally, Figure 2.27 shows the effects of varying the separation d_t between the elements on the top layer. The separation is varied from 2 mm to 6 mm with $l_t = l_b = 33$ mm, and $h = 7$ mm. It can be seen that the frequency corresponding to the lower edge of the bandwidth increases with increasing d_t. Also, as d_t increases, the upper resonance disappears, leading to a considerable decline in the bandwidth. Optimum impedance matching can be achieved when the patch elements are separated at around 4 mm.

Therefore, the size of the top patches, the spacing between the dielectric layers as well as the separation between the top patches collectively affect the impedance matching of the two resonances. The upper resonance is controlled predominantly by the size of the top patches, and the size of the bottom patch primarily controls the lower resonance.

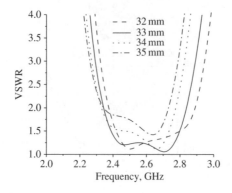

Figure 2.25 VSWR for various sizes of top patches l_t. (Reproduced by permission of John Wiley & Sons, Inc.[69])

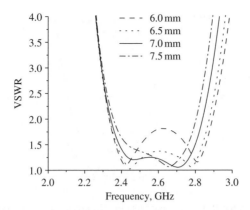

Figure 2.26 VSWR for various spacings h. (Reproduced by permission of John Wiley & Sons, Inc.[69])

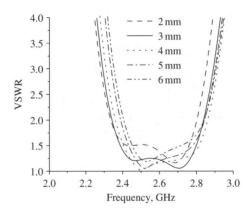

Figure 2.27 VSWR for various separations d_t. (Reproduced by permission of John Wiley & Sons, Inc.[69])

Figure 2.28 shows the measured VSWR for the optimized antenna configuration. The measured 1.5:1 VSWR bandwidth reaches 16% with a center frequency of 2.45 GHz, covering the whole Industrial, Scientific and Medical (ISM) band well.

The radiation patterns were then measured at three primary frequencies, namely 2.40 GHz, 2.45 GHz, and 2.50 GHz. Figures 2.29 and 2.30 display the far-field radiation patterns in the y–z and x–z planes, respectively. Each of the radiation patterns (E_θ and E_ϕ) is normalized by the maxima for all the three frequencies in both x–z and y–z planes. In the planes of interest, the radiation patterns for the E_θ and E_ϕ components are found to be relatively stable and symmetrical, with absence of side lobes in the upper-half space. In the x–z plane, the average half-power beamwidths for E_θ/E_ϕ components are 54°/61° and 62°/53° in the y–z plane. Owing to the finite-size ground plane used in the measurement, the front-to-back

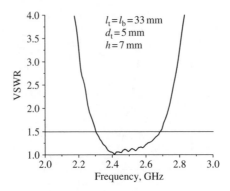

Figure 2.28 Measured VSWR for optimal design. (Reproduced by permission of John Wiley & Sons, Inc.[69])

(a)

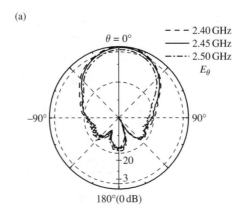

(b)

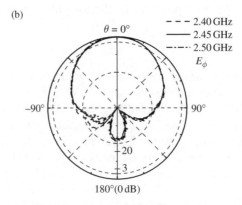

Figure 2.29 Measured radiation patterns for optimal design in the y–z plane. (Reproduced by permission of John Wiley & Sons, Inc.[69])

(a)

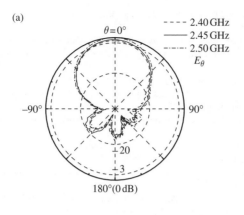

(b)

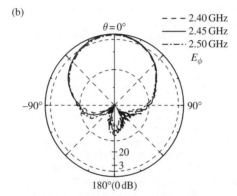

Figure 2.30 Measured radiation patterns for optimal design in the x–z plane. (Reproduced by permission of John Wiley & Sons, Inc.[69])

ratios are generally lower than -20 dB in both planes. In addition, the measured average gain in both planes is approximately 9.5 dBi.

Therefore, this broadband 45°-linear polarized stacked antenna covers more than the required 1.5:1 VSWR bandwidth for WLAN applications in the 2.45-GHz ISM band. This allows the antenna to be installed in a varying environment. Such a design also features stable, symmetrical and 45°-linear polarization characteristics with the absence of side lobes, low front-to-back radiation levels, and high gain across the operating band. Another distinct advantage is that the antenna can be constructed simply without the need for a complicated feeding network.

REFERENCES

[1] I. J. Bahl and P. Bhartia, *Microstrip Antennas*. Dedham, MA: Artech House, 1980.

[2] J. R. James, P. S. Hall and C. Wood, *Microstrip Antennas: Theory and Design*. London: Peter Peregrinus (IEE), 1981.

[3] K. C. Gupta and A. Benalla, *Microstrip Antenna Design*. Norwood, MA: Artech House, 1988.

[4] J. R. James and P. S. Hall, *Handbook of Microstrip Antennas*. London: Peter Peregrinus (IEE), 1988.

[5] P. Bhartia, K. V. S. Rao and R. S. Tomar, *Millimeter-wave Microstrip and Printed Circuit Antennas*. Boston, MA: Artech House, 1991.

[6] Y. T. Lo and E. S. W. Lee, *Antenna Handbook*. New York: Van Nostrand Reinhold, 1993.

[7] D. M. Pozar and D. H. S. (Ed.), *Microstrip Antennas: Analysis and Design*. New York: John Wiley & Sons, Inc., 1995.

[8] K. F. Lee, K. M. Luk, K. F. Tong, S. M. Shum, T. Huynh and R. Q. Lee, 'Experimental and simulation studies of the coaxially fed U-slot rectangular patch antenna,' *IEE Proceedings: Microwave, Antennas and Propagation*, vol. 144, no. 5, pp. 354–358, 1997.

[9] R. Garg, P. Bhartia, I. Bahl and A. Ittipiboon, *Microstrip Antenna Design Handbook*. Boston, MA: Artech House, 2001.

[10] R. E. Munson, 'Conformal microstrip antennas and microstrip phased arrays,' *IEEE Transactions on Antennas and Propagation*, vol. 22, no. 1, pp. 74–78, 1974.

[11] J. W. Howell, 'Microstrip antennas,' *IEEE Transactions on Antennas and Propagation*, vol. 23, no. 1, pp. 90–93, 1975.

[12] J. R. James, P. S. Hall, C. Wood and A. Henderson, 'Microstrip antennas,' *IEEE Transactions on Antennas and Propagation*, vol. 29, no. 1, pp. 124–128, 1981.

[13] K. R. Carver and J. W. Mink, 'Microstrip antenna technology,' *IEEE Transactions on Antennas and Propagation*, vol. 29, no. 1, pp. 2–24, 1981.

[14] D. M. Pozar, 'Microstrip antennas,' *IEEE Transactions on Antennas and Propagation*, vol. 40, no. 1, pp. 79–81, 1992.

[15] J. P. Daniel, G. Dubost, C. Terret, J. Citerne and M. Drissi, 'Microstrip antennas,' *IEEE Antennas and Propagation Magazine*, vol. 35, no. 1, pp. 14–38, 1993.

[16] I. Bahl, P. Bhartia and S. Stuchly, 'Design of microstrip antennas covered with a dielectric layer,' *IEEE Transactions on Antennas and Propagation*, vol. 29, no. 3, pp. 314–318, 1981.

[17] M. D. Deshpande and M. C. Bailey, 'Input impedance of microstrip antennas,' *IEEE Transactions on Antennas and Propagation*, vol. 30, no. 4, pp. 645–650, 1982.

[18] D. M. Pozar, 'Input impedance and mutual coupling of rectangular microstrip antennas,' *IEEE Transactions on Antennas and Propagation*, vol. 30, no. 6, pp. 1191–1196, 1982.

[19] J. L. Cao, W. L. Dong and W. X. Li, 'Radiation field of pentagonal microstrip antenna,' *IEEE Transactions on Antennas and Propagation*, vol. 34, no. 1, pp. 103–106, 1986.

[20] M. D. Deshpande and N. K. Das, 'Rectangular microstrip antenna for circular polarization,' *IEEE Transactions on Antennas and Propagation*, vol. 34, no. 5, pp. 744–746, 1986.

[21] V. Palanisamy and R. Garg, 'Analysis of arbitrarily shaped microstrip patch antennas using segmentation technique and cavity model,' *IEEE Transactions on Antennas and Propagation*, vol. 34, no. 10, pp. 1208–1213, 1986.

[22] R. G. Vaughan, 'Two-port higher mode circular microstrip antennas,' *IEEE Transactions on Antennas and Propagation*, vol. 36, no. 3, pp. 309–321, 1988.

[23] R. Kastner, E. Heyman and A. Sabban, 'Spectral domain iterative analysis of single- and double-layered microstrip antennas using the conjugate gradient algorithm,' *IEEE Transactions on Antennas and Propagation*, vol. 36, no. 9, pp. 1204–1212, 1988.

[24] M. A. Sultan, 'The mode features of an ideal-gap open-ring microstrip antenna,' *IEEE Transactions on Antennas and Propagation*, vol. 37, no. 2, pp. 137–142, 1989.

[25] M. Himdi, J. P. Daniel and C. Terret, 'Analysis of aperture-coupled microstrip antenna using cavity method,' *Electronics Letters*, vol. 25, no. 6, pp. 391–392, 1989.

[26] R. Q. Lee, T. Huynh and K. F. Lee, 'Experimental study of the cross-polarization characteristics of rectangular patch antennas,' *IEEE International Symposium on Antennas and Propagation*, vol. 2, pp. 624–627, June 1989.

[27] N. Jayasundere and T. S. M. Maclean, 'Omnidirectional radiation patterns from body-mounted microstrip antennas,' *IEE Sixth International Conference on Antennas and Propagation*, vol. 1, pp. 187–190, 4–7 April 1989.

[28] E. Levine, G. Malamud, S. Shtrikman and D. Treves, 'A study of microstrip array antennas with feed network,' *IEEE Transactions on Antennas and Propagation*, vol. 37, no. 4, pp. 426–434, 1989.

[29] A. E. Gera, 'The radiation resistance of a microstrip element,' *IEEE Transactions on Antennas and Propagation*, vol. 38, no. 4, pp. 568–570, 1990.

[30] Z. Nie, W. C. Chew and Y. T. Lo, 'Analysis of the annular-ring-loaded circular-disk microstrip antenna,' *IEEE Transactions on Antennas and Propagation*, vol. 38, no. 6, pp. 806–813, 1990.

[31] D. Thouroude, M. Himdi and J. P. Daniel, 'CAD-oriented cavity model for rectangular patches,' *Electronics Letters*, vol. 26, no. 13, pp. 842–844, 1990.

[32] D. M. Pozar and B. Kaufman, 'Design considerations for low sidelobe microstrip arrays,' *IEEE Transactions on Antennas and Propagation*, vol. 38, no. 8, pp. 1176–1185, 1990.

[33] G. P. Gauthier, A. Courtay and G. M. Rebeiz, 'Microstrip antennas on synthesized low dielectric-constant substrates,' *IEEE Transactions on Antennas and Propagation*, vol. 45, no. 8, pp. 1310–1314, 1997.

[34] M. I. Aksun, S. L. Chuang and Y. T. Lo, 'On slot-coupled microstrip antennas and their applications to cp operation: theory and experiment,' *IEEE Transactions on Antennas and Propagation*, vol. 38, no. 8, pp. 1224–1230, 1990.

[35] D. R. Jackson and N. G. Alexopoulos, 'Simple approximate formulas for input resistance, bandwidth and efficiency of a resonant rectangular patch,' *IEEE Transactions on Antennas and Propagation*, vol. 39, no. 3, pp. 407–410, 1991.

[36] N. K. Das and D. M. Pozar, 'Multiport scattering analysis of general multilayered printed antennas fed by multiple feed ports. Part II: Applications,' *IEEE Transactions on Antennas and Propagation*, vol. 40, no. 5, pp. 482–491, 1992.

[37] J. R. M. S. A. Bokhari and F. E. Gardiol, 'Radiation pattern computation of microstrip antennas on finite-size ground planes,' *IEE Proceedings: Microwave, Antennas and Propagation*, vol. 139, no. 3, pp. 278–286, 1992.

[38] D. M. Pozar, 'Radiation and scattering characteristics of microstrip antennas on normally biased ferrite substrates,' *IEEE Transactions on Antennas and Propagation*, vol. 40, no. 9, pp. 1084–1092, 1992.

[39] D. M. Pozar, 'Microstrip antennas and arrays on chiral substrates,' *IEEE Transactions on Antennas and Propagation*, vol. 40, no. 10, pp. 1260–1263, 1992.

[40] S. D. Targonski and D. M. Pozar, 'Design of wideband circularly polarized aperture-coupled microstrip antennas,' *IEEE Transactions on Antennas and Propagation*, vol. 41, no. 2, pp. 214–220, 1993.

[41] T. Kashiwa, T. Onishi and I. Fukai, 'Analysis of microstrip antennas on a curved surface using the conformal grids FD–TD method,' *IEEE Transactions on Antennas and Propagation*, vol. 42, no. 3, pp. 423–427, 1994.

[42] P. S. Hall and I. L. Morrow, 'Analysis of radiation from active microstrip antennas,' *IEE Proceedings: Microwave, Antennas and Propagation*, vol. 141, no. 5, pp. 359–366, 1994.

[43] L. Giauffret, J. M. Laheurte and A. Papiernik, 'Experimental and theoretical investigations of new compact large bandwidth aperture-coupled microstrip antenna,' *Electronics Letters*, vol. 31, no. 16, pp. 2139–2140, 1995.

[44] L. Giauffret and J. M. Laheurt, 'Theoretical and experimental characterisation of CPW-fed microstrip antennas,' *IEE Proceedings: Microwave, Antennas and Propagation*, vol. 143, no. 1, pp. 354–358, 1996.

[45] H. Iwasaki, 'A circularly polarized small-size microstrip antenna with a cross slot,' *IEEE Transactions on Antennas and Propagation*, vol. 44, no. 10, pp. 1399–1401, 1996.

[46] S. K. Satpathy, K. P. Ray and G. Kumar, 'Compact shorted variations of circular microstrip antennas,' *Electronics Letters*, vol. 34, no. 2, pp. 137–138, 1998.

[47] H. M. An, B. K. J. C. Nauwelaers and A. R. van de Capelle, 'Broadband microstrip antenna design with the simplified real frequency technique,' *IEEE Transactions on Antennas and Propagation*, vol. 42, no. 2, pp. 129–136, 1994.

[48] U. K. Revankar and A. Kumar, 'Broadband stacked three-layer circular microstrip antenna arrays,' *Electronics Letters*, vol. 28, no. 21, pp. 1995–1997, 1992.

[49] H. M. An, B. K. J. C. Nauwelaers and A. R. van de Capelle, 'Broadband circularly polarised microstrip antenna in two-sided structure with coaxial probe coupling,' *Electronics Letters*, vol. 29, no.3, pp. 310–312, 1993.

[50] J. M. Carroll, K. A. Tilley, S. Kanamaluru and K. Chang, 'Slot coupling of coplanar waveguide to patch antennas suitable for MMIC applications,' *Electronics Letters*, vol. 30, no.15, pp. 1195–1196, 1994.

[51] U. K. Revankar and A. Kumar, 'Mutual coupling between stacked three-layer circular microstrip antenna elements,' *Electronics Letters*, vol. 30, no. 24, pp. 1997–1998, 1994.

[52] H. Iwasaki, 'A circularly polarized rectangular microstrip antenna using single-fed proximity-coupled method,' *IEEE Transactions on Antennas and Propagation*, vol. 43, no. 8, pp. 895–897, 1995.

[53] S. Egashira and E. Nishiyama, 'Stacked microstrip antenna with wide bandwidth and high gain,' *IEEE Transactions on Antennas and Propagation*, vol. 44, no. 10, pp. 1533–1534, 1996.

[54] M. Clenet and L. Shafai, 'Wideband single-layer microstrip antenna for array applications,' *Electronics Letters*, vol. 35, no. 2, pp. 1292–1293, 1999.

[55] S. M. Duffy, 'An enhanced bandwidth design technique for electromagnetically coupled microstrip antennas,' *IEEE Transactions on Antennas and Propagation*, vol. 48, no. 2, pp. 161–164, 2000.

[56] M. A. G. de Aza, J. Zapata and J. A. Encinar, 'Broad-band cavity-backed and capacitively probe-fed microstrip patch arrays,' *IEEE Transactions on Antennas and Propagation*, vol. 48, no. 7, pp. 784–789, 2000.

[57] M. A. Khayat, J. T. Williams and S. A. L. D. R. Jackson, 'Mutual coupling between reduced surface-wave microstrip antennas,' *IEEE Transactions on Antennas and Propagation*, vol. 48, no. 10, pp. 1581–1593, 2000.

[58] S. Hudson and M. D. Pozar, 'Grounded coplanar waveguide-fed aperture-coupled cavity-backed microstrip antenna,' *Electronics Letters*, vol. 36, no. 12, pp. 1003–1005, 2000.

[59] B. Lee and F. J. Harackiewicz, 'Miniature microstrip antenna with a partially filled high-permittivity substrate,' *IEEE Transactions on Antennas and Propagation*, vol. 50, no. 8, pp. 1160–1162, 2002.

[60] K. P. Ray, G. Kumar and H. C. Lodwal, 'Hybrid-coupled broadband triangular microstrip antennas,' *IEEE Transactions on Antennas and Propagation*, vol. 51, no. 1, pp. 139–141, 2003.

[61] R. Sauleau and P. Coquet, 'Influence of residual air gaps on characteristics of circularly polarised aperture-coupled millimetre-wave microstrip antennas,' *Electronics Letters*, vol. 39, no. 12, pp. 889–891, 2003.

[62] J. T. Bernhard, E. Kiely and G. Washington, 'A smart mechanically actuated two-layer electromagnetically coupled microstrip antenna with variable frequency, bandwidth, and antenna gain,' *IEEE Transactions on Antennas and Propagation*, vol. 49, no. 4, pp. 597–601, 2001.

[63] C. Wood, 'Improved bandwidth of microstrip antennas using parasitic elements,' *IEE Proceedings: Microwave, Antennas and Propagation*, vol. 127, no. 4, pp. 231–234, 1980.

[64] G. Kumar and K. C. Gupta, 'Broad-band microstrip antennas using additional resonators gap-coupled to the radiating edges,' *IEEE Transactions on Antennas and Propagation*, vol. 32, no. 12, pp. 1375–1379, 1984.

[65] G. Kumar and K. C. Gupta, 'Nonradiating edges and four edges gap-coupled multiple resonator broadband microstrip antennas,' *IEEE Transactions on Antennas and Propagation*, vol. 33, no. 2, pp. 173–178, 1985.

[66] C. K. Aanandan, P. Mohanan and K. G. Nair, 'Broadband gap coupled microstrip antenna,' *IEEE Transactions on Antennas and Propagation*, vol. 38, no. 10, pp. 1581–1586, 1990.

[67] Y. Lubin and A. Hessel, 'Wide-band, wide-angle microstrip stacked-patch-element phased arrays,' *IEEE Transactions on Antennas and Propagation*, vol. 39, no. 8, pp. 1062–1070, 1991.

[68] J. T. Aberle, D. M. Pozar and J. Manges, 'Phased arrays of probe-fed stacked microstrip patches,' *IEEE Transactions on Antennas and Propagation*, vol. 42, no. 7, pp. 920–927, 1994.

[69] T. S. P. See and Z. N. Chen, 'Design of dual-polarization stacked arrays for ISM band applications,' *Microwave and Optical Technology Letters*, vol. 38, no. 2, pp. 142–147, 2003.

3

Broadband Suspended Plate Antennas

3.1 INTRODUCTION

Many techniques have been developed to broaden the impedance bandwidth of single-element single-layer microstrip patch antennas as discussed in Chapter 2. The use of thick dielectric substrates is a simple but effective method to enhance the impedance bandwidth of a microstrip patch antenna by reducing its unloaded Q. However, it is important to note that as the impedance bandwidth increases, surface wave losses also increase owing to the thick dielectric substrate which reduces the radiation efficiency. A practical method to suppress the surface waves is to lower the permittivity of the substrate. The medium between the planar radiator and ground plane may be a low-permittivity or air-filled substrate such as foam ($\epsilon_r \approx 1.07$), or even air. As described in Chapter 1, so-called suspended plate antennas (SPAs) with thicknesses ranging from $0.03\lambda_1$ to $0.12\lambda_1$ (λ_1 is the wavelength corresponding to the lower edge of the well-matched impedance bandwidth) and a low relative dielectric constant of ~ 1 have a broad impedance bandwidth and unique radiation performance. SPAs give rise to special design considerations, although they are essentially a variation of microstrip patch antennas defined by the IEEE Standard 145-1993: a thin metallic conductor bonded to a thin grounded dielectric substrate.

SPAs are not difficult to fabricate. The necessary elements include a ground plane, a radiating plate, a feeding structure, and low-permittivity supports if necessary. Figure 3.1 shows the geometry. Usually, an inexpensive foam layer or air is used between the radiating plate and the ground plane, and the spacing h between the plate and the ground plane ranges from $0.03\lambda_1$ to $0.12\lambda_1$. A coaxial probe or probe-like feeding structure is often used to excite the plate because of the large spacing and the low dielectric constant. However, the long

Broadband Planar Antennas: Design and Applications   Zhi Ning Chen and Michael Y. W. Chia
© 2006 John Wiley & Sons, Ltd

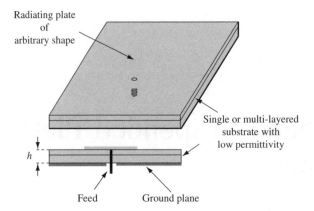

Figure 3.1   Geometry of a suspended plate antenna.

probe can result in poor impedance matching owing to the large feed inductance. In addition, the impedance bandwidth of an SPA is limited to about 8 % for VSWR = 2, due to the poor impedance matching as illustrated in Figure 1.5, as the spacing is larger than about 0.06 times the operating wavelength. Therefore, it is necessary to develop impedance-matching techniques to attain impedance bandwidths of more than 8 % for SPAs.

In this chapter, techniques for obtaining broad impedance bandwidths are introduced by means of several practical designs. First, compensation techniques using a capacitive load are reviewed. Cutting slots from the radiators is then introduced as one of the most important broadband techniques as for microstrip patch antennas. Then, the use of electromagnetic coupling between the feeding probe and SPA to neutralize the large feed inductance within a broad frequency range is elaborated. Next, a broadband SPA with a center-concaved radiator is designed and compared with a conventional SPA. Last, the insertion of an impedance transition between the feeding probe and plate is described, with a design example of an SPA with a vertical feeding sheet. The key to achieving a broadband impedance response of an SPA is to compensate for the high inductance or/and resistance between the feeding probe and the radiating plate. In contrast to the techniques outlined in Chapter 2, the matching network or impedance transition can be inserted under the radiating plate or in the same plane.[1,2]

The chapter goes on to describe techniques to enhance the radiation performance. A dual-probe feeding arrangement consisting of a feed probe and a capacitive load is introduced with a design example. Then, a feeding scheme with a half-wavelength feeding strip is presented. Next, a center-fed SPA with a symmetrical shorting pin and an SPA with a pair of symmetrically arranged feed probe and shorting pin are discussed. After that, a broadband SPA fed by double L-shaped probes is designed. Last, the functions of shorting pins and slots applied in SPAs are discussed.

The chapter then goes on to discuss arrays with nonplanar ground planes and suspended plate elements in a compact design. The mutual coupling between the suspended elements installed on the nonplanar ground planes is first investigated. Then, arrays with nonplanar ground planes are designed and compared.

3.2 TECHNIQUES TO BROADEN IMPEDANCE BANDWIDTH

3.2.1 CAPACITIVE LOAD

The significant inductance introduced by a long coaxial probe can be neutralized by introducing capacitance. Some techniques have been developed as shown in Figure 2.19, where the dielectric substrate is replaced with a low-permittivity material such as a foam layer of $\epsilon_r \approx 1.07$ or air for a broad bandwidth in excess of 20%.[3-6] The capacitance is inserted between the feed probe and the radiator to compensate for the large inductance across the broad bandwidth.

An SPA with a Rectangular Loop Slot

Consider the example in Figure 3.2. The SPA comprises a 70 mm × 70 mm square patch adhering to a dielectric layer with $\epsilon_r = 3.38$ and 60-mil thickness (about 1.52 mm), and is excited by a 50-Ω coaxial probe of radius 0.6 mm and length 15 mm. The medium between the dielectric layer and the ground plane is air. The use of a thin dielectric has two primary merits. One is that the patch can be etched on the dielectric layer more easily than the fabrication of a PEC (such as copper) plate. The other merit is that the dielectric layer can increase the coupling between the feed and the radiator, while hardly narrowing the impedance bandwidth. The SPA resonates at about 1.6 GHz, where the return loss reaches a minimum, and its overall electrical thickness is about $0.08\lambda_0$ at 1.6 GHz.

   Figures 3.3(a) and (b) show that the inductive impedance of the feed probe for the SPA without any slot is too large to be matched to the feed structure. In order to offset the large input inductance, a rectangular annular slot is etched around the feed point on the patch. The slot length is fixed at 10 mm and the slot width d is varied. By comparing the simulated input impedance with and without the loop slot as shown in Figures 3.3(a) and (b), it can be seen that with the capacitive slot, the large feed resistance with a peak value of 150 Ω decreases to about 90 Ω, while the large inductive reactance is significantly reduced to around zero in the resonant regions. The wider the slot, the larger is the capacitance.

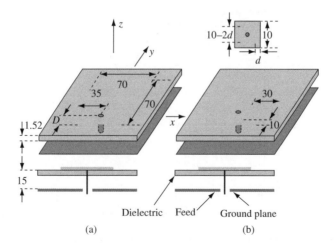

Figure 3.2 SPAs (a) without loop slot; (b) with loop slot (dimensions in millimeters).

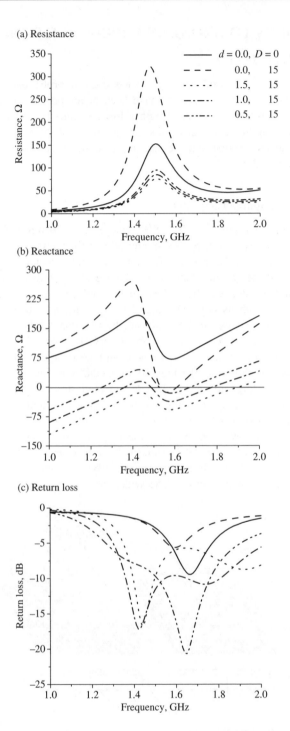

Figure 3.3 Input impedance and return loss for the SPA with capacitive load (dimensions in millimeters).

Figure 3.3(c) shows the impedance matching. By optimizing the slot width, the impedance bandwidth for $|S_{11}| < -10\,\text{dB}$ can reach up to 30 %.

3.2.2 SLOTTED PLATES

The introduction of additional resonators is often used to widen the impedance bandwidth of a microstrip patch antenna. These parasitic resonators (usually patches) can be added on the same layer as the main patch or stacked above the main patch, as described in Chapter 2. The former arrangement introduces additional resonance by gap-coupling but increases the lateral size of the antenna. In the stacked arrangement, by means of strong coupling between the main patch and stacked patches, the additional resonances are well excited. For single-layered SPAs, in contrast, a possible way to increase the bandwidth is to cut one slot or multiple slots of various shapes on the radiating plate. These parasitic slots will not increase the volume but are intended to introduce additional resonances.

Figure 3.4(a) depicts a single-layered single-element SPA as reported by Huynh and Lee in 1995.[7] A long U-shaped slot was cut symmetrically from the plate. The slot effectively offsets the large input inductance due to the long probe, so that the SPA was able to attain an impedance bandwidth of between 10 % and 40 % using a single layer and without any parasitic elements. A lot of effort has been devoted to refining this approach.[8–15] These

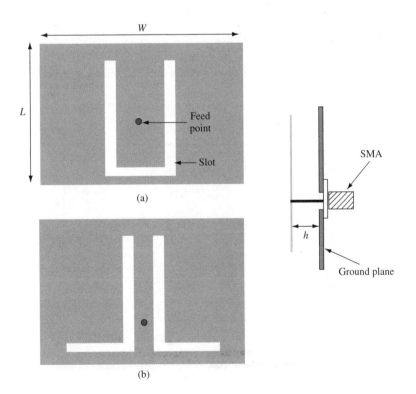

Figure 3.4 SPAs with (a) U-shaped slot, and (b) double-L slots.

studies have suggested that a large aspect ratio W/L and a thickness of $h > 0.06\lambda_1$ are essential to attain an impedance bandwidth of 20–40 %, where λ_1 represents the wavelength corresponding to the lower edge of the bandwidth. Also, it has been established that the long U-slot cut from the plate introduces an additional resonance adjacent to the original one, so that the bandwidth of a long U-slot SPA can reach to 20–40 %.[11,12]

Slots of other shapes may be used. For example, two L-shaped slots cut symmetrically on the plate were used to achieve a broad impedance bandwidth as shown in Figure 3.4(b).[16,17] An impedance bandwidth of more than 22 % for VSWR $= 2$ with slight enhancement of radiation performance has been demonstrated.

Many studies have shown that cutting long slots of various shapes on the radiator can improve the impedance matching between the feed probe and the suspended plate over a broad frequency range of up to 30 %. The impedance matching is sensitive to the size, location and shape of the slots.

3.2.3 CASE STUDY: SPA WITH AN Ω-SHAPED SLOT

Consider the SPA shown in Figure 3.5.[17,18,19] The plate has a width of 64 mm and a length L and is suspended at a height of 10 mm. An Ω-shaped slot is cut symmetrically from the plate with respect to the y-axis. A 50-Ω probe with a radius of 0.6 mm excites the plate close to its midpoint. By keeping the rest of the geometrical parameters constant, the effect of varying L has been examined, the length being varied from 64 mm to 122 mm.

Figure 3.6 illustrates the return loss, which show broad impedance bandwidths for $L \geq 72$ mm. For example, the SPA with $L = 72$ mm has a 22 % bandwidth from 1.78 GHz to 2.2 GHz for $|S_{11}|$ less than -10 dB. The larger L is, the better the impedance matching condition. In particular, when $L = 112$ mm, good matching occurs at a higher frequency

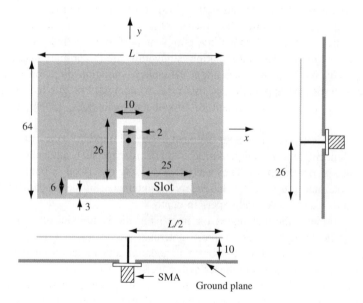

Figure 3.5 Geometry of an SPA with an Ω-shaped slot (dimensions in millimeters).

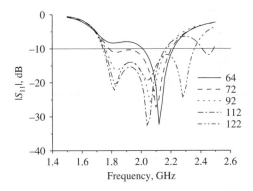

Figure 3.6 $|S_{11}|$ for SPAs with various lengths L (dimensions in millimeters).

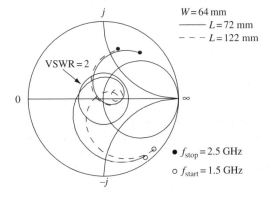

Figure 3.7 Input impedance of SPAs with $L = 72$ mm and 122 mm.

region of 2.45 GHz. As L is increased to about 122 mm, the bandwidth reaches up to 30 %, ranging from 1.75 GHz to 2.36 GHz.

Figure 3.7 shows the input impedance loci for $L = 72$ mm and $L = 122$ mm. The smaller SPA has a single looped impedance locus. The small loop is located around the center of the Smith chart and within the VSWR = 2 circle. The impedance locus for $L = 122$ mm features two small loops around the center of the Smith chart, resulting in the wide bandwidth.

The input impedance is shown in Figure 3.8. It is important to note that a second parallel resonance appears at $L = 112$ mm. Good impedance matching is shown at this resonance for $L = 122$ mm. Varying the length L slightly affects the lower resonance.

To understand the radiation characteristics at the resonances, it is useful to identify the modes at which the antenna is operating. Thus, the radiation patterns for SPAs with different lengths L were studied at several typical frequencies over the bandwidth of interest, such as the lower and upper edges of the bandwidth, as well as at some frequencies within the bandwidth. Figure 3.9 shows the radiation patterns for $L = 72$ mm at 1.78, 1.84, 2.10 and 2.21 GHz. The cross-polarized radiation in the E plane is not given because it is much lower than the co-polarized radiation.

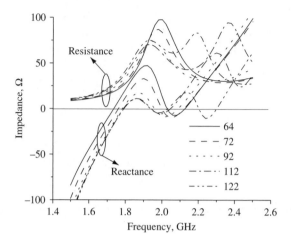

Figure 3.8 Input impedance of SPAs with varying L (dimensions in millimeters).

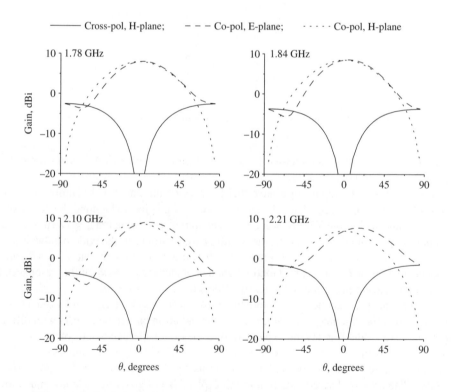

Figure 3.9 Radiation patterns for SPAs with $L = 72$ mm at various frequencies.

Three important points can be made. The first point is that the SPA operates in the dominant TM_{01} mode. The second point is that, in the H-planes, the co-pol and cross-pol radiation patterns are symmetrical with respect to the E-plane over the bandwidth. The co-to-cross-pol ratios in the H-plane vary from 12.5 dB at 2.10 GHz to 9.2 dB at 2.21 GHz. The third point is that the co-pol radiation patterns in the E-plane are somewhat asymmetrical about the H-planes owing to the asymmetrical and slotted structures. At the higher frequencies, the radiation patterns are squinted 10° to 15° away from the boresight. However, it should be noted that the co-to-cross-pol ratios in the H-planes are higher than 18 dB within the half-power beamwidths. In addition, the maximum gain ranges from 7.5 dBi to 8.9 dBi over the bandwidth.

Figure 3.10 shows the radiation patterns for the SPA with $L = 122$ mm at 1.82, 2.04, 2.28 and 2.36 GHz. Besides having the same impedance features as the SPA with $L = 72$ mm, the SPA with $L = 122$ mm has the obvious distinction in its radiation patterns. The cross-pol radiation levels in the H-planes are much higher than those of the SPA with $L = 72$ mm even within the half-power beamwidths. With increasing frequency, the cross-pol radiation levels in the H-planes rapidly increase whereas the co-pol radiation levels decrease. In particular, the cross-pol radiation levels exceed the co-pol radiation levels by 8 dB at 2.28 GHz owing to the undesired higher-order mode TM_{20} near 2.1 GHz. Therefore, the SPA has different radiation performance at the higher frequencies. In addition, owing to excitation of the higher-order modes, the achieved gain at the boresight significantly decreases, even lower than 0 dBi. Therefore, for applications requiring maximum boresight radiation and/or polarization purity,

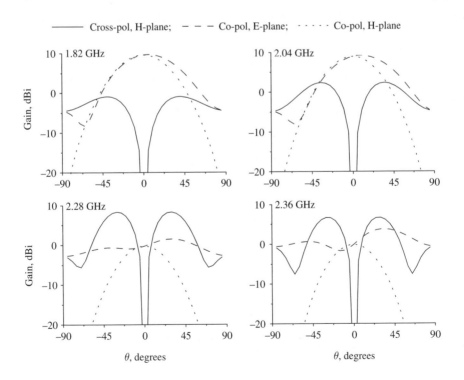

Figure 3.10   Radiation patterns for the SPA with $L = 122$ mm at various frequencies.

the usable operating bandwidth, considering both impedance and radiation bandwidths, is only 22 % instead of 30 %.

3.2.4 ELECTROMAGNETIC COUPLING

Electromagnetic (EM) coupling feed techniques are often used with the microstrip patch antennas, to increase the bandwidths. As described in Chapter 2, proximity, stacked and aperture structures are all typical EM coupling designs. In microstrip patch antennas, the thin and high-permittivity dielectric substrates make it possible to transfer RF and microwave energy from transmission structures to radiators effectively by the strong EM coupling. However, because of the large spacing and low-permittivity medium between ground plane and radiator, the aforementioned techniques cannot be applied directly to SPAs. Probes or probe-like structures are more suitable for SPAs, hence much effort has been devoted to the modification of existing probes or the construction of new probe structures. This section reviews the application of L-shaped and T-shaped probes in broadband SPA designs.

L-shaped Probe

In 1997, Nakano and co-workers described a curl antenna with an L-shaped probe electromagnetically coupled to a wire radiator.[20] By dint of the EM coupling between the L-shaped feed probe and the radiating curl, the curl antenna attained a broadband impedance response. In 1998, Luk and co-workers used the L-shaped probe to excite a rectangular SPA, as shown in Figure 3.11.[21] Unlike the conventional probe-fed structures, the probe consists of an additional horizontal portion, which is placed under the radiating plate, instead of directly connecting the probe to the radiator. The gap between the horizontal portion and the plate is so small that the plate can be effectively excited by the strong EM coupling. The impedance bandwidth for VSWR = 2 reaches 36 %. This structure has been modified and widely applied in array design for cellular base stations[22].

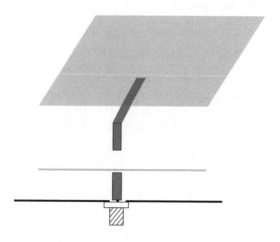

Figure 3.11 Geometry of an SPA fed by an L-probe.

T-shaped Probe

A T-shaped probe can also be used to achieve a broadband SPA. By using the EM coupling between the probe-fed T-shaped strip and the suspended plate, the large input inductance due to the long probe can effectively be compensated for and a broad bandwidth obtained.[23,24] Compared with the L-shaped probe, the T-shaped probe features a simpler mechanical structure with a co-planar configuration.

As an example, a rectangular SPA excited by a probe-fed strip has been investigated experimentally. The geometry is illustrated in Figure 3.12. The aspect ratio (width to length) of the plate should be chosen to suppress the excitation of higher-order modes. Two designs were accomplished with the geometrical dimensions listed in Table 3.1.

For convenience, the rectangular plate consists of a patch etched on to a rectangular printed-circuit board (PCB) of the same size as the patch. The dielectric substrate of the PCB has a thickness of $t = 16$ mil (about 0.4 mm) and a dielectric constant of $\epsilon_r = 3.38$. The patch is suspended at a height h, parallel to the ground plane. The narrow strip ($l \times w$) fed by a 50-Ω coaxial probe is electromagnetically coupled and on the same plane as the patch. The vertical probe feeds the strip at location ($L/2$, d). In contrast to the slotted plate antenna, an electrically narrow notch of width $s = 1$ mm is etched on to the patch with the longer sides parallel to the non-radiating edges. This co-planar (single-layer) structure offers

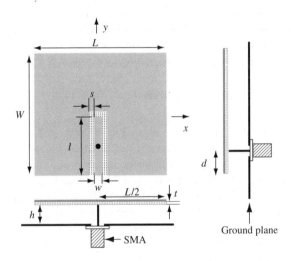

Figure 3.12 Geometry of an SPA fed by a T-shaped probe. (Reproduced by permission of IEE.[23])

Table 3.1 Parameters of a T-proble-fed SPA.

Antenna	$W \times L$ (mm^2)	h (mm)	$l \times w$ (mm^2)	d (mm)
1	66×72	13	30×2	14
2	62×72	14	16×7	1

a mechanical advantage over the two-layer designs and plays a similar role to that of the L-shaped probe as described earlier.

A square plate driven directly by a coaxial probe is first designed to operate at frequency f_o, and is suspended at a height of about 0.1 λ_o. The plate is notched to form a narrow probe-fed strip strongly coupled to the patch. By changing the size of the strip and the location of the feed point, the matching condition can be controlled. The dimensions of the plate affect both the resonant frequency and radiation performance. Usually, the excitation of undesirable higher-order modes that significantly contribute to the cross-polarized radiation at higher frequencies can be suppressed when the aspect ratio is close to unity. Good matching can be achieved near the desired operating frequency by adjusting the dimensions of the strip and/or the plate. The plate of the final design may be of any shape.

For the measurements, a rectangular, perfectly electrically conducting (PEC) plate of 310 mm×460 mm was used to approximate the infinite ground plane. The measured impedance bandwidths for VSWR = 2 are 26 % (1.64 GHz to 2.12 GHz) for Antenna 1 and 36 % (1.59 GHz to 2.29 GHz) for Antenna 2. It is manifest that, owing to the EM coupling, the impedance bandwidth has been widened appreciably. The EM coupling between the strip and the patch acts as an impedance matching network, which effectively cancels out the large reactance stemming from the long probe. In the design of Antenna 1, a narrow strip was used and good matching could be attained by adjusting the location of the probe. For Antenna 2, as the probe feed is located at the strip edge, the length and width of the strip could be changed to obtain good matching.

In addition, a parameter known as the *matching factor* (MF) is used to assess the matching property of broadband antennas[25]. Mathematically, the matching factor represents an average VSWR of a broadband antenna within the bandwidth for a certain VSWR value (e.g. VSWR = 2) and describes its impedance matching performance. It is calculated from

$$\mathrm{MF}_{\mathrm{VSWR}=2} = \mathrm{VSWR}_0 - \frac{\Delta A_{\mathrm{VSWR}=2}}{\Delta f_{\mathrm{VSWR}=2}}. \tag{3.1}$$

The term $\Delta A_{\mathrm{VSWR}=2}$ denotes the area between the VSWR ≤ 2 curve and the relevant $\mathrm{VSWR}_0 = 2$ line, and $\Delta f_{\mathrm{VSWR}=2}$ is the frequency range within which the VSWR remains less than 2. In fact, the $\mathrm{MF}_{\mathrm{VSWR}=2}$ describes the matching condition in the frequency range for VSWR = 2. By using equation 3.1, Antennas 1 and 2 have matching factors of 1.28 and 1.70, respectively. From the definition of the MF, a smaller MF means a better matching condition.

Figure 3.13 demonstrates a VSWR curve corresponding to good impedance matching. For applications requiring an impedance bandwidth for VSWR = 1.5, Antenna 2 is a better option than Antenna 1. However, to meet a demand of >30 % bandwidth for VSWR = 2, Antenna 1 should be selected.

The radiation patterns of Antenna 1 at the center frequency of 1.85 GHz are plotted in Figure 3.14. The cross-pol radiation levels and half-power beamwidths are given in Table 3.2. It is clear that the co-to-cross-pol ratios in the E-plane are typically higher than 25 dB but lower in the H-plane; i.e. from 17 dB to 22 dB. The half-power beamwidths are stable for both E- and H-planes, with a maximum difference of less than 4°. In addition, it is clear that the antenna with an aspect ratio L/W of 1.125 is able to suppress the excitation of higher-order TM_{nm} modes effectively. The use of the narrow strip prevents severe distortion of the current distribution on the plate surface. The strip structure also alleviates the impact of the higher-order modes (TM_{0m}) on the cross-polarized radiation. Therefore, this feeding

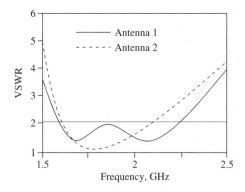

Figure 3.13 Measured VSWR of the SPA fed by a T-shaped probe. (Reproduced by permission of IEE.[23])

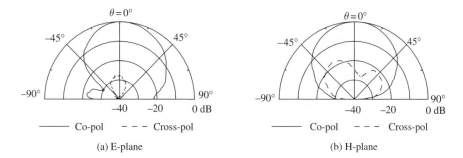

(a) E-plane (b) H-plane

Figure 3.14 Measured radiation patterns for the SPA fed by a T-shaped probe at 1.85 GHz in E- and H-planes. (Reproduced by permission of IEE.[23])

Table 3.2 Measured co-to-cross-pol ratios and half-power beamwidths for the T-probe-fed SPA.

f (GHz)	1.65	1.75	1.85	2.0	2.10
Co-to-cross-pol ratio(E/H-plane) (dB)	25/22	25/21	27/21	32/17	29/17
Half-power beamwidth(E/H-plane) (deg)	48/54	48/54	48/52	50/52	52/56

structure not only simplifies the mechanical design to some extent but also ameliorates the radiation performance.

3.2.5 NONPLANAR PLATES

The radiator of a conventional SPA is a planar plate installed parallel to the ground plane. For the single-element SPA, factors affecting the impedance bandwidth usually include the radiator shape, the feed structure and thickness. Therefore, as compared with the conventional microstrip patch antenna, the design of the SPA actually has one more degree of freedom arising from the spacing between the radiating plate and the ground plane.

Nonplanar microstrip patch antennas can be used to reduce the size of radiating patches. For example, a microstrip patch antenna with an elevated center portion has a 53 % reduction in size along its E-planes with an increase in its impedance bandwidth.[26] However, the height of the antenna is greatly increased by up to a quarter of the patch length. The nonplanar patch can be used to increase the radiation efficiency of a microstrip patch antenna. By elevating the radiating edges of the patch, the gain increased by about 3 dBi with a 50 % increase in the patch length and more than double the height.[27]

On the other hand, it is possible to improve the impedance bandwidth of an SPA without any increase in its volume, by changing the conventional planar radiating plate to a non-planar structure.[28] The non-planar radiating plate makes full use of the space between the radiating plate and the ground plane. The center portion of the SPA is caved in so that it is close to the ground plane, whereas the main section of the plate is suspended at a larger height above the ground plane. Figure 3.15 shows the geometry.

The parameters of this SPA have been determined by simulation. The plate parallel to the x–y plane measures 70 mm along both axes. The spacing between the plate and the ground plane is $h_1 = 15$ mm. The center portion of the plate is a 'V' shape, and the distance between its bottom and the ground plane is $h_2 = 2$ mm. The upper part of the 'V' is 6 mm in width and the bottom part is 3 mm in width. A coaxial probe feeds the plate at a point $(x = 0$ mm, $y = -19$ mm, $z = 2$ mm$)$ or $d = 16$ mm at the bottom of the 'V' through a 50-Ω SMA connector.

For comparison, two planar SPAs of size 70 mm×70 mm have been fabricated. The heights of the planar SPAs were, respectively, $h = 2$ mm and $h = 15$ mm. The feed points were located at $(x = 0$ mm, $y = -12$ mm, $z = 2$ mm$)$ and $(x = 0$ mm, $y = -34$ mm, $z = 15$ mm$)$, respectively, for optimal impedance matching. In addition, a copper plate measuring

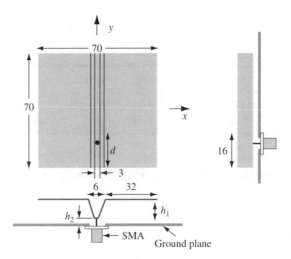

Figure 3.15 Geometry of the SPA with a concave center portion (dimensions in millimeters). (Reproduced by permission of IEEE.[28])

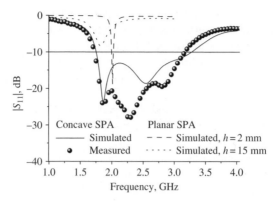

Figure 3.16 Comparison of measured and simulated return losses of the SPA with a concave center portion. (Reproduced by permission of IEEE.[28])

310 mm × 225 mm was used in the tests instead of the infinite ground plane assumed in the simulations.

Figure 3.16 compares the bandwidths for $|S_{11}| < -10\,\text{dB}$ of 58 % (measured) and 61 % (simulated). For planar SPAs, the SPA with a height of $h = h_1 = h_2 = 2\,\text{mm}$ achieves a bandwidth of only 2 % due to its higher Q. The impedance matching for the SPA with a height of $h = h_1 = h_2 = 15\,\text{mm}$ is poor due to the large input inductance caused by the long probe, although the location of the feed point is changed. Therefore, the large height of the concaved SPA lowers the Q factor of the SPA, and the concaved center portion of the plate compensates for the large inductance over a broad bandwidth in the order of 60 %.

Figure 3.17 shows input impedances. For the planar SPA with a small spacing $h = 2\,\text{mm}$, the input resistance and reactance are low within the frequency range of 1 GHz to 4 GHz.

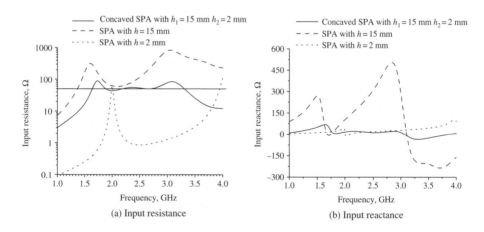

Figure 3.17 Input impedances of the SPA with a concave center portion. (Reproduced by permission of IEEE.[28])

The planar SPA with $h = 15$ mm has a high input resistance and inductive reactance. For the concave SPA, the resistance varies around 50 Ω, and the reactance around 0 Ω. Owing to the concave center portion of the SPA, the high inductance has been significantly cancelled out. Moreover, it is evident that there are three resonances within the bandwidth as an additional resonance has been well excited around 2.27 GHz because of the concave portion.

Figures 3.18(a) and (b) compare the radiation patterns in E- and H-planes at 1.87, 2.27 and 2.82 GHz. This comparison shows that the characteristics of the co-polarized and cross-polarized radiation patterns at 1.87 GHz and 2.82 GHz are typical of conventional SPAs. The minimum co-to-cross-pol ratios in the H-planes for $-90° < \theta < 90°$ are around 8 dB within the half-power beamwidths, although the radiation patterns at 2.27 GHz have the higher cross-polarized radiation levels around $\pm45°$. The high cross-polarized radiation levels degrade the polarization characteristics and may not be suitable for applications requiring high polarization purity. Moreover, the measured average gain at the boresight is about 8 dBi across the bandwidth.

It is evident that a probe-fed suspended antenna with a concave center portion is capable of providing a broad impedance bandwidth of up to 60 % for $|S_{11}| < -10$ dB due to the additional mode. Furthermore, compared with the microstrip patch antenna, the spacing between the radiating plate and the ground plane provides an additional design parameter, which can be used to enhance the performance of the SPA (e.g. the impedance bandwidth). However, the drawback of this method is the degraded radiation performance.

3.2.6 VERTICAL FEED SHEET

Another technique to improve the impedance bandwidth of an SPA is to make use of the spacing between the radiating plate and the ground plane by situating the feed probe under the radiating plate. For instance, instead of using a thin cylindrical probe feed structure, a copper sheet fed by a coaxial probe is connected to the radiating plate.[29] This not only

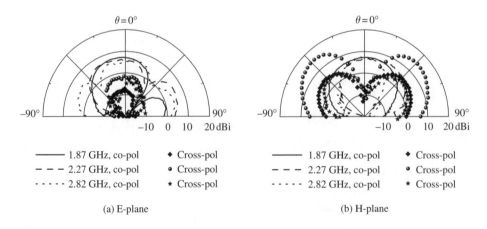

Figure 3.18   Measured radiation patterns for an SPA with a concave center portion at 1.87, 2.27 and 2.82 GHz in E- and H-planes. (Reproduced by permission of IEEE.[28])

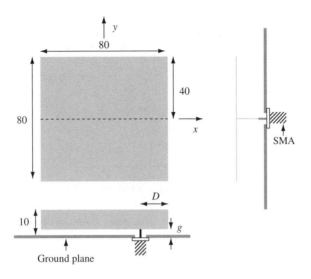

Figure 3.19 An SPA with a vertical feed sheet (dimensions in millimeters).

increases the thickness of the feed probe but also allows the radiating plate to be fed by the sheet along a line, whereas the plate is fed at a point with a thin feed probe.

As an example, a square SPA was designed and its geometry shown in Figure 3.19, where the coordinate system is oriented so that the ground plane measuring 310 mm × 310 mm lies in the x–y plane and its z-axis is centered vertically through the ground plane. The sides of the plate are parallel to the x- and y-axes, respectively. This design employs a conventional square plate antenna operating at 2 GHz.

An 80 mm × 80 mm square plate is placed concentrically above the ground plane at a height 10 mm. A rectangular copper sheet with dimensions 9.5 mm × 80 mm is used as an impedance transformer between the feed probe and the plate. It is centrally and vertically connected to the plate along the x-axis, and fed by a 50-Ω coaxial probe at its bottom. The distance between the feed point and the sheet edge is indicated by D, and the feed gap between the bottom of the sheet and the ground plane by g. As a result, the plate is excited by the probe-fed sheet in a line, not at a point. The impedance matching can be improved by adjusting the electromagnetic coupling between the sheet bottom and the ground plane, the location of the feed point, the shape of the radiator. By adjusting the distance D and the gap g, good impedance matching for VSWR $= 2$ can readily be achieved, when the distance D is around 12 mm and the gap g varies from 0.5 mm to 1 mm. The input capacitance is greatly increased as the gap becomes larger. Figure 3.20 illustrates the measured VSWR for the $D = 12.5$ mm and $g = 0.5$ mm. The SPA has achieved a 2:1 VSWR impedance bandwidth of 60 %, which covers the frequency range of 1.58 GHz to 2.96 GHz.

Figures 3.21 and 3.22 show the measured radiation patterns for both E_θ and E_ϕ components, in the x–z and y–z planes. For the E_θ components, the SPA operates at a patch-like dominant mode for operating frequencies lower than about 2 GHz. Above 2 GHz, a deep notch appears in the radiation patterns for the E_θ components owing to the dominance of the higher-order modes. The back radiation levels for the E_θ components and radiation levels for the E_ϕ components increase significantly. Also, Figures 3.21 and 3.22 demonstrate that the

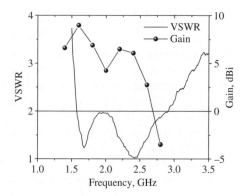

Figure 3.20 Measured VSWR and gain of an SPA with a vertical feed sheet. (Reproduced by permission of John Wiley & Sons, Inc.[29])

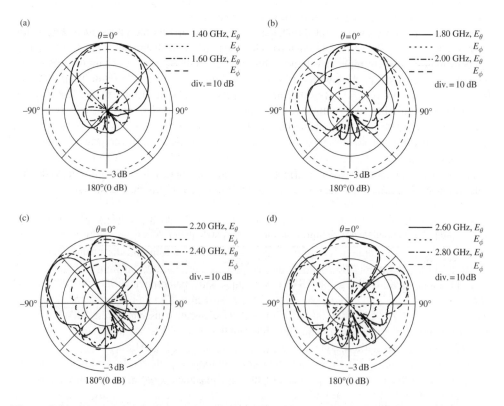

Figure 3.21 Measured radiation patterns for the SPA with a vertical feed sheet at 1.40, 1.60, 1.80, 2.00, 2.20, 2.40, 2.60 and 2.80 GHz in the x–z plane. (Reproduced by permission of John Wiley & Sons, Inc.[29])

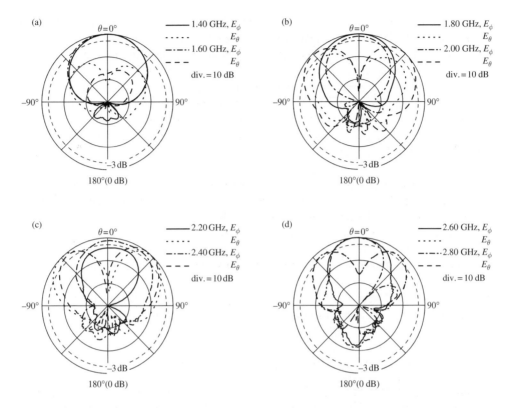

Figure 3.22 Measured radiation patterns for the SPA with a vertical feed sheet at 1.40, 1.60, 1.80, 2.00, 2.20, 2.40, 2.60 and 2.80 GHz in the y–z plane. (Reproduced by permission of John Wiley & Sons, Inc.[29])

maximum radiation of the E_θ components occurs at around the boresight ($\theta = 0°$) across the bandwidth, although the radiation patterns are severely distorted at the higher frequencies.

The measured maximum gain is plotted in Figure 3.20. It has been found that at the lower frequencies, ranging from 1.6 GHz to 2.0 GHz pertaining to the dominant mode, the gain is greater than 5 dBi. However, the gain varies between 2.7 dBi and 6.5 dBi within the frequency range of 2.0 GHz to 2.6 GHz owing to the severe distortion in the radiation patterns.

Therefore, by making good use of the large spacing between the radiator and ground plane, the feed structure can be modified to attain good impedance matching across a broad bandwidth owing to the well-matched adjacent higher-order mode. However, the radiation performance of the SPA, such as radiation pattern, gain, maximum radiation direction and back radiation level, might vary with the frequency, in particular at the higher frequencies.

3.3 TECHNIQUES TO ENHANCE RADIATION PERFORMANCE

The radiation performance of an antenna is a crucial design consideration for applications requiring pure polarization, such as base-station arrays used for cellular wireless communication systems.

Earlier sections of this chapter have demonstrated the radiation patterns for several SPAs across well-matched bandwidths. However, all the SPAs suffered from undesirable changes in their radiation characteristics within the achieved impedance bandwidths, especially at the higher operating frequencies. Typically, the co-polarized radiation patterns in the E-plane are distorted, with an asymmetric deep dip at the higher frequencies. This distortion also results to some degree in a squint of the radiation patterns. The co-to-cross-pol ratio in the H-plane decreases significantly owing to the increase in cross-polarized radiation levels, resulting in low gain and high interference between systems with diverse polarizations. This degradation in radiation performance limits their application, especially in systems requiring polarization purity.

Many techniques have been developed over the years to alleviate the degraded radiation performance. Dual-feed probes, a half-wavelength feeding strip and a probe-fed center slot are effective methods for single-element SPAs or SPA arrays.

This section of the chapter discusses the radiation characteristics of probe-fed SPAs. The co-polarized and cross-polarized radiation patterns for SPAs fed by a conventional probe and a non-radiating probe are compared.[30] After that, techniques to improve radiation performance are introduced and some examples provided to validate the ideas. The concept of a symmetrical, balanced and center-fed feeding scheme is derived. This has been shown to be conducive for the enhancement of radiation performance of broadband SPAs.

3.3.1 RADIATION CHARACTERISTICS

For comparison, non-radiating feed structures – modified transmission structures – have been explored. The radiation characteristics of SPAs excited by a non-radiating feed structure and a conventional probe-driven strip (probe) have been compared experimentally.[31–33]

Non-radiating Probe-driven Strips

To study the effects of radiation from a long feed probe on the radiation characteristics of probe-fed SPAs, probe-driven feed structures without any radiation have been designed. The two types of feed structure are shown in Figure 3.23. Each comprises a narrow metallic strip with dimensions $S \times w_f$ (40 mm $\times$4 mm) in the y–z plane, which is etched on to a dielectric slab (Roger4003, $\epsilon_r = 3.38$) symmetrically along the z-axis and driven at its bottom by a 50-Ω coaxial probe through an SMA. The dielectric slab with dimensions $S \times W \times t$ (40 mm $\times$40 mm $\times$1.52 mm) is vertically set above a ground plane and grounded at its bottom. Of the two structures, Feed 1 is a transmission-line-like structure, whereby the probe-driven strip is backed by the grounded second strip of size $S \times w$. Feed 2 is a type of co-planar-waveguide-like structure, whereby the probe-driven strip is in the same plane as the strips of dimensions $S \times w/2$. Both strips are grounded at the bottom and separated from the probe-driven strip by a gap $d = 1.0$ mm.

For Feed 1, its E_θ components were measured in the x–y (horizontal) plane. The testing frequencies corresponded to the minima of the measured return losses, which vary from 1.70 GHz to 1.79 GHz. Figure 3.24 shows the received power for different widths w (0–40 mm), which has been normalized by the maximum at $w = 0$ mm. With increasing

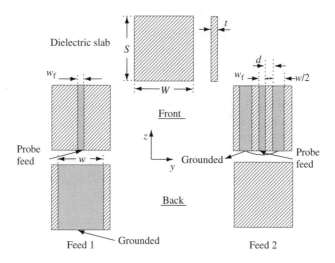

Figure 3.23 Geometry of two probe-driven strips. (Reproduced by permission of IEEE.[30])

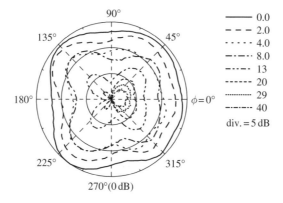

Figure 3.24 E_θ component radiation patterns of Feed 1 with varying width, w (in millimeters). (Reproduced by permission of IEEE.[30])

width w, the grounded strip acts like a vertical ground plane. As a result, the received power becomes weaker and the directivity of the radiation patterns increases gradually.

Similarly, Feed 2 was tested at frequencies ranging from 1.84 GHz to 2.0 GHz. Figure 3.25 shows the radiation patterns of the E_θ components in the x–y plane as the widths ($w/2$) of the grounded strips increased from 0 mm to 17 mm. The results clearly show that the larger the width $w/2$, the smaller the received power. However, unlike Feed 1, Feed 2 has nearly omnidirectional radiation patterns for all the widths w.

Figure 3.26 plots the maximum radiated power from Feeds 1 and 2 against the width ratio w/w_f, which are normalized by the power radiated from the probe-driven strip without any grounded strip ($w = 0$ mm). It is seen that the radiated power is reduced by 3 dB when the width ratios for both feed structures increase to about 1.5. When both width ratios approach

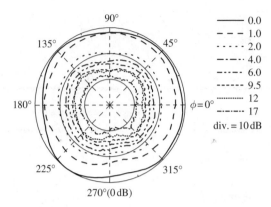

Figure 3.25 E_θ component radiation patterns of Feed 2 with varying width, $w/2$ (in millimeters). (Reproduced by permission of IEEE.[30])

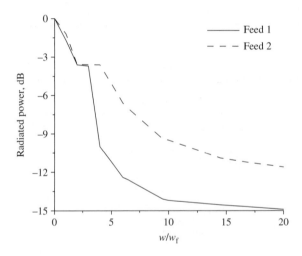

Figure 3.26 Measured radiated power of Feed 1 and 2 against the width ratio of w/w_f. (Reproduced by permission of IEEE.[30])

10, the radiated power drops to -14.2 dB and -10.6 dB for Feeds 1 and 2, respectively. In other words, the power radiated from the feed strips with width ratio 10 is only 3.8 % and 9 % of those from the probe-driven strip without ground ($w = 0$ mm).

Fundamentally, the electric currents on the vertical strips can be divided into two components, namely the transmission-line mode of equal magnitude and a 180° phase shift, as well as the monopole mode of the same phase but with unequal magnitude. The ratio of the two components is predominantly controlled by the width ratio as well as the thickness and dielectric constant of the slab. In particular, the probe-driven strip operates as a conventional monopole backed by a dielectric slab as w approaches zero. When the width ratio w/w_f is large, Feed 1 acts as a vertical microstrip transmission line, and Feed 2 functions as a vertical co-planar waveguide, which are open-circuited at their top ends. Thus, the radiation from the currents on both the vertical ground plane and probe-driven strip of Feed 1 or 2 is almost cancelled out.

Furthermore, the measured input impedance illustrated in Figure 3.27 shows that the feed structures with wide vertical ground planes can be considered as non-radiating due to their high Q, operating as a vertical microstrip transmission line and a vertical co-planar waveguide, respectively.

SPAs with and without Probe Radiation

Figure 3.28 shows the geometry of two rectangular SPAs with and without probe radiation. The copper plate measuring $W \times L$ (70 mm $\times$ 70 mm) is parallel to a ground plane and placed at a height of $H = 10$ mm. The ground plane is a 320 mm $\times$ 285 mm perfectly electrically conducting (PEC) plate. A probe-driven strip similar to Feed 1 (above) with dimensions $H \times w_f$ (10 mm $\times$ 2 mm) was used as a feed structure, and etched on to a 10 mm $\times$ 20 mm $\times$ 0.813 mm dielectric slab. Antenna 1 was fed by a probe-driven strip without a vertical

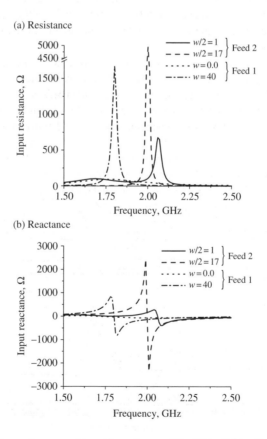

Figure 3.27 Comparison of measured input impedance of the SPAs fed by Feed 1 and 2 (dimensions in millimeters). (Reproduced by permission of IEEE.[30])

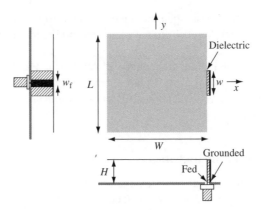

Figure 3.28 Geometry of the suspended plate antenna with a probe-type feed. (Reproduced by permission of IEEE.[30])

ground plane, and Antenna 2 with a vertical ground plane of size $H \times w$ (10 mm $\times$ 20 mm). Both the antennas were fed centrally at the edge of the plates. From the discussions in the previous section, Antennas 1 and 2 can be considered here to be with and without probe radiation, respectively.

Figure 3.29 shows that the resonant frequency for Antenna 2 is 2.22 GHz with $W/\lambda = 0.518$, which is higher than the 1.91 GHz for Antenna 1 with $W/\lambda = 0.446$, although their radiating plates are of the same size. The achieved impedance bandwidths for VSWR $= 2$ are 7.3 % and 9.9 % for Antennas 1 and 2, respectively. In order to keep the resonant frequency of Antenna 2 close to that of Antenna 1, the plate length of Antenna 2 is extended to $W = 78$ mm. Therefore, instead of Antenna 2, Antenna 3 with dimensions $L \times W \times H$ (70 mm $\times$ 78 mm $\times$ 10 mm) will be taken into consideration in the following study. Figure 3.29 shows that the achieved 2:1 VSWR bandwidth of Antenna 3 is 8.5 %. Antenna 3 can be compared to Antenna 1 because the extended plate length of Antenna 3 is almost equal to the overall length of the original plate (70 mm) plus the probe-driven strip length (10 mm).

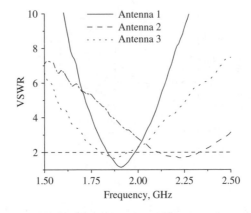

Figure 3.29 Measured VSWR of three SPAs. (Reproduced by permission of IEEE.[30])

Within the measured impedance bandwidths, the radiation patterns for Antennas 1 and 3 will be examined in three principal planes, namely x–z (E-plane), y–z (H-plane), and $x - y$. Both the co-polarized and cross-polarized radiation patterns were measured at 1.84, 1.91 and 1.98 GHz for Antenna 1 and at 1.80, 1.88, and 1.96 GHz for Antenna 3. These frequencies correspond to the lower edge, center and upper edge of the bandwidth.

Figure 3.30 compares the radiation patterns in the E-plane. Across their entire bandwidths, the shapes of the radiation patterns for Antennas 1 and 3 are nearly the same. The co-polarized

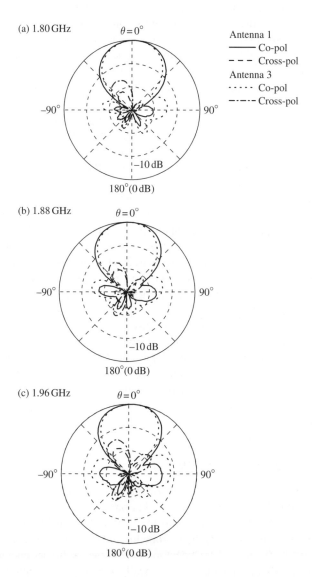

Figure 3.30 Measured co- and cross-polarized radiation patterns in the E-plane. (Reproduced by permission of IEEE.[30])

half-power beams are about 50° in width and symmetrical with respect to the boresight ($\theta = 0°$). Each of the co-polarized radiation patterns is distorted with two asymmetrical side lobes close to the horizon ($\theta = \pm 90°$). Moreover, the co-to-cross-pol ratios for Antennas 1 and 3, respectively, vary from 21 dB to 18 dB and from 17 dB to 14 dB with increasing frequency. The cross-polarized radiation of Antenna 3 is about 3 dB higher than that of Antenna 1, and the co-polarized radiation patterns of Antenna 3 with larger side lobes have more severe distortion than those of Antenna 1, although the probe radiation in Antenna 3 has been suppressed effectively.

Figure 3.31 shows that Antennas 1 and 3 have almost the same radiation features in the H-plane. The co-polarized radiation patterns are symmetrical with respect to the boresight, and the half-power beamwidths are about 74° across the bandwidth. As frequency increases, the co-to-cross-pol ratios for Antennas 1 and 3, respectively, vary from 14 dB to 11 dB and from 7.5 dB to 5 dB. The dip in the cross-polarized radiation patterns are lower than −20 dB around the boresight. The co-to-cross-pol ratios for Antenna 3 are much lower than those for Antenna 1.

In addition, the measured radiation patterns of the E_θ and E_ϕ components in the x–y plane further prove the absence of probe radiation from Antenna 3. Figure 3.32 displays the radiation patterns for Antennas 1 and 3, which have been normalized by the maximum E_θ components. The shapes of the radiation patterns of the E_θ and E_ϕ components remain nearly the same across the bandwidth. At the higher frequencies, the radiated power for the E_ϕ components is low but higher for the E_θ components. Comparing Figures 3.32(a) and (b), the radiation patterns of Antenna 1 are more sensitive to frequency than those of Antenna 3. This is because the radiation of the E_θ components is derived from both the monopole mode of the feed and the cavity mode of the plate antenna. Radiation from the monopole mode (the probe radiation) is more sensitive to the frequency. This suggests that the probe radiation from Antenna 3 has been suppressed effectively. On the other hand, the radiation patterns in Figures 3.32(c) and (d) confirm that the horizontally polarized fields are derived primarily from the plate rather than the feed of Antenna 1 or 3. Also, the cross-polarized radiation is always very low along $\phi = 0°$ and 180° directions.

Therefore, the radiation performance of Antenna 3 has not been improved despite the significant suppression of probe radiation. This could suggest that probe radiation from the feeds used in suspended plate antennas does not contribute greatly to the increase in cross-polarized radiation in the H-plane and the distortion of co-polarized radiation patterns in E-plane, although the radiation in the horizontal (x–y) plane well retains its omnidirectionality.

3.3.2 SPA WITH DUAL FEED PROBES

In order to study the degraded radiation performance of probe-fed suspended plate antennas, Figure 3.33 schematically compares the distributions of the imaginary currents induced on the plates at the higher frequency within the achieved broad impedance bandwidth. A conventional probe-fed SPA is driven by a probe located close to one of the radiating edges. It has an asymmetrical structure with respect to the H-plane. At higher operating frequencies within an operating bandwidth, the plate is electrically longer than the length for the operating dominant mode. As a result, the higher-order modes are gradually excited so that the induced current distribution on the plate is distorted. The distorted current distribution at the plate

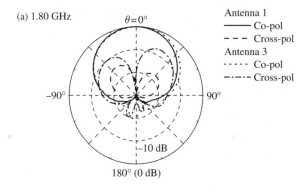

(a) 1.80 GHz

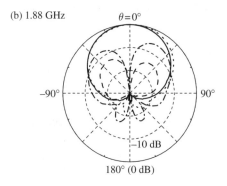

(b) 1.88 GHz

(c) 1.96 GHz

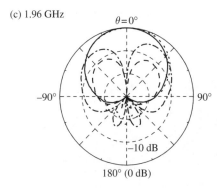

Figure 3.31 Measured co- and cross-polarized radiation patterns in the H-plane. (Reproduced by permission of IEEE.[30])

has the cross-polarized current components around the feed point. These cross-polarized components radiate the cross-polarized fields.

In the E-plane, the radiated cross-polarized fields are cancelled out completely owing to the symmetrical but out-of-phase cross-polarized currents. In the H-plane, however,

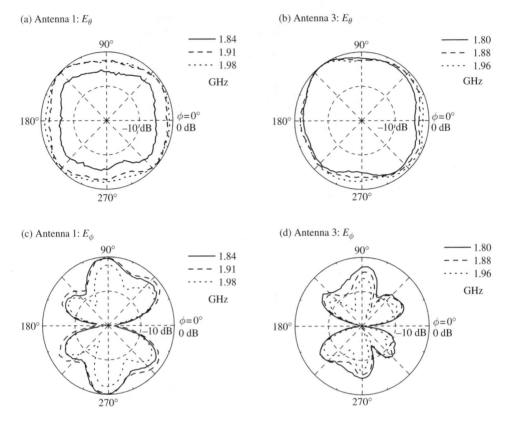

Figure 3.32 Measured co- and cross-polarized radiation patterns in the x–y plane. (Reproduced by permission of IEEE.[30])

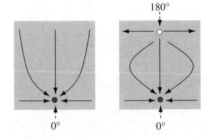

Figure 3.33 Distributions of induced electric currents on the plates for single- and dual-probe schemes.(Reproduced by permission of IEEE.[30])

the radiated cross-polarized fields cannot becancelled out completely in the cuts off the boresight. Moreover, the co-polarized current components in the E-plane are asymmetrical with respect to the H-plane owing to the excitation of the higher-order modes, which cause the asymmetrical co-polarized radiation in the E-plane.

Therefore, excitation of the higher-order modes contributes significantly to the increase in cross-polarized radiation levels in the H-plane and the asymmetry of the co-polarized radiation in the E-plane. For an SPA having a broad bandwidth of more than 6%, it is difficult to avoid the excitation of higher-order modes.[35] To alleviate the impairment of the higher-order modes on the radiation performance of the broadband SPA, one additional probe carrying out-of-phase currents with the same amplitude is located symmetrically on the opposite side, as illustrated in Figure 3.33.[33,36-38]

The second out-of-phase probe excites the same currents as those by the original probe. However, for the second probe, the co-polarized currents in the E-plane are in phase. The cross-polarized currents in the H-plane are out of phase with respect to both the E- and H-planes. Therefore, the two feed probes form a structurally symmetrical and electrically balanced configuration. More importantly, characteristics of the dual-probe feed structure are frequency independent. With the symmetrical and balanced arrangement, the cross-polarized radiation in the H-plane can be cancelled out, and the co-polarized radiation patterns in the E-plane can be made symmetrical with respect to the H-plane. SPAs fed by the dual probes can suppress the cross-polarized radiation and improve the co-polarized radiation patterns, although the undesired higher-order modes may not be suppressed completely.

However, this technique also suffers from some limitations. For example, it needs a broadband network in order to keep the currents on the two feed probes out of phase within a broad operating bandwidth for single-element designs. The second problem is that two feed probes are required. This problem may be alleviated when the dual-probe technique is used in array environments.[39] Also, the cross-polarized radiation levels cannot be suppressed greatly in the cuts off the principal E- and H-planes, owing to the phase difference due to the difference in the distances of the two feed points with respect to a far-field observation point.

3.3.3 CASE STUDY: CENTER-CONCAVED SPA WITH DUAL FEED PROBES

As an example, a center-concaved SPA excited by dual probes will be examined. This antenna excited by a single probe has been discussed in section 3.2.5 (as shown in Figure 3.15). The earlier analysis showed that this type of SPA is capable of providing an impedance bandwidth of 60% for a −10 dB return loss, but its radiation performance was severely degraded, as demonstrated in Figure 3.18.

The dual-probe technique will be used to mitigate the radiation performance of this SPA. A second 50-Ω probe is added on the opposite side of the first probe, as show in Figure 3.34. The two feed points are symmetrical with respect to the x-axis. The currents on the two identical probes have the same magnitude but are out of phase.

Computer simulations suggest that the antennas with single-probe and dual-probe feeding structures have almost the same impedance matching conditions, as shown in Figure 3.35. The two probes in the dual-probe structure have the same return loss response although they are out of phase. The slight change in the response for the two out-of-phase probes may be due to the introduction of the second probe, which can be considered as a 50-Ω resistive loading of the first one, and vice versa.

For comparison, the radiation patterns are plotted in Figure 3.36. For the single probe, the co-polarized radiation patterns in the E-plane is asymmetrical owing to the asymmetrical

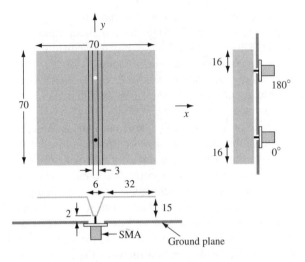

Figure 3.34 Center-concaved SPA fed by dual feeding probes (dimensions in millimeters).

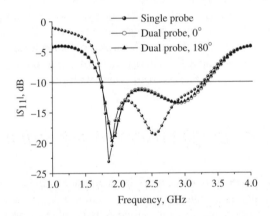

Figure 3.35 Comparison of return losses of SPAs fed by single and dual probes.

structure of the SPA even at the lower frequency of 1.8 GHz, whereas in the H-plane they are symmetrical with respect to the E-plane. The cross-polarized radiation patterns in the E-plane are very much lower owing to the cancellation of the symmetrical but out-of-phase cross-polarized currents at the plate. The same phenomenon is observed in the SPA fed by dual probes. However, the cross-polarized radiation levels in the H-planes become high when the observation point is away from the boresight for the single-probe antenna. With the introduction of the second out-of-phase probe, a structurally symmetrical and electrically balanced SPA is formed. Theoretically, the cross-polarized radiation is completely cancelled out such that the cross-polarized radiation levels for the dual-probe case are much lower than the co-polarized radiation levels. It can be concluded that the introduction of the second probe can alleviate the degraded radiation performance.

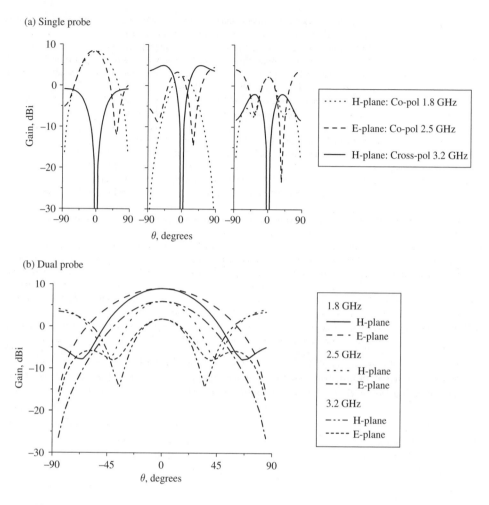

Figure 3.36 Comparison of radiation patterns for SPAs fed by single and dual probes.

3.3.4 SPA WITH HALF-WAVELENGTH PROBE-FED STRIP

A single-probe feed structure comprising a probe-fed half-wavelength strip can be employed as an alternative to enhance the radiation performance of broadband SPAs.[40] This can result in a simple as well as broadband design.

The geometry is shown in Figure 3.37. The SPA operates around 1.8 GHz with an operating wavelength of $\lambda_o = 167$ mm in free space. The rectangular, thin brass copper plate measuring 66 mm × 72 mm is placed parallel to a ground plane. The spacing between the radiating plate and the ground plane is 10 mm ($\sim 0.06\lambda_o$). The medium between the plate and ground plane is air. The perfectly electrically conducting (PEC) plate with dimensions 305 mm × 225 mm is used to approximate the infinite ground plane. A narrow conducting strip with dimensions 8 mm × 68 mm is suspended horizontally between the plate and the ground plane. The separation between the strip and the ground plane, as well as between the strip and the plate,

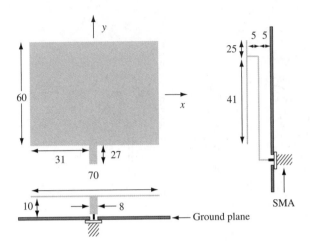

Figure 3.37 Geometry of an SPA fed by a half-wavelength probe-driven strip (dimensions in millimeters). (Reproduced by permission of IEEE.[40])

is 5 mm. The strip is symmetrically located along the midline (y-axis) of the plate. One of the strip ends is excited by a 50-Ω coaxial probe extended from an SMA connector. The other end is connected to the radiating plate. The currents on the two vertical segments have an approximate 180° phase shift due to the $0.47\lambda_o$ long strip connected at the two ends. By properly locating the feed point on the radiating plate, the impedance matching between the probe-fed strip and the SMA connector can be controlled. Basically, the two vertical segments form a balanced feeding structure to achieve cancellation of the cross-polarized radiation in both E- and H-planes. The probe-fed strip forms a broadband impedance matching network through the electromagnetic coupling between the plate and the strip.

The measured input impedance is plotted against frequency in Figure 3.38. It is seen that between two resonances at 1.70 GHz and 1.95 GHz, the input resistance varies around

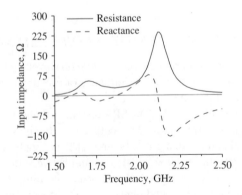

Figure 3.38 Measured input impedance of the SPA fed by a half-wavelength strip. (Reproduced by permission of IEEE.[40])

50 Ω, and the input reactance is around zero. The large input reactance resulting from the long probe has been effectively neutralized due to the strong electromagnetic coupling, which acts as an impedance matching network. In addition, the shortened vertical segments are also good for suppression of the input reactance. The measured VSWR against frequency shown in Figure 3.39 exhibits that the impedance bandwidth for VSWR = 2 reaches 19.5 % (1.62–1.97 GHz), which greatly exceeds the 8 % bandwidth for the conventional probe-fed SPA.

The co- and cross-polarized radiation patterns in the E- and H-planes (y–z and x–z planes) were measured at 1.63, 1.80 and 1.95 GHz. Figure 3.40 reveals the measured radiation patterns in the E-planes, where the co-to-cross-polarized ratios are higher than 20 dB at all frequencies and in all directions. Furthermore, the half-power beamwidths for the co-polarized radiation patterns vary from 54° to 74° across the bandwidth. At frequencies 1.80 GHz and 1.95 GHz, dips of 20 dB appear in the co-polarized radiation patterns near $\theta = 45°$. This distortion of the co-polarized radiation pattern has been observed also in other single-probe-fed SPAs and is mainly due to the asymmetrical structure and excitation of the higher-order modes. For this SPA, the probe-fed feed strip asymmetrically excites the radiating plate. At the higher frequencies, the co-pol current distribution on the radiating plate is asymmetrical with respect to the x-axis, which is undesirable.

Similarly, the radiation patterns in the H-plane are illustrated in Figure 3.41. At all frequencies, the measured co-to-cross-pol ratios are more than 20 dB. Compared with the slit cross-polarized radiation patterns for conventional probe-fed SPAs, each of the cross-polarized radiation patterns has only one narrow beam in the upper half space. Also, the co-polarized radiation patterns are unchanged with the half-power beamwidths ranging from 64° to 68° and symmetrical with respect to the z-axis. In contrast to conventional probe-fed SPAs, this SPA features a significant improvement in the radiation performance owing to the suppression of undesirable cross-polarized radiation. In particular, this improvement has been accomplished over the 20 % impedance bandwidth. Moreover, the average gain of 5 dBi was measured across the 20 % impedance bandwidth as plotted in Figure 3.42, and was lowered mainly due to the finite-size ground plane.

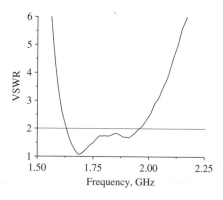

Figure 3.39 Measured VSWR of the SPA fed by a half-wavelength strip. (Reproduced by permission of IEEE.[40])

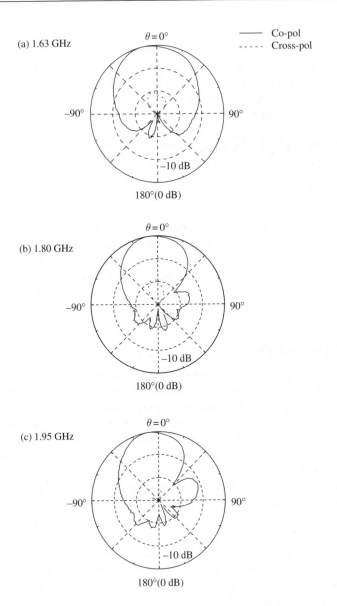

Figure 3.40 Measured co- and cross-polarized radiation patterns in the E-plane. (Reproduced by permission of IEEE.[40])

   Compared to the design analyzed above, an SPA fed by a wider feeding strip has a smaller impedance bandwidth and almost the same cross-polarized radiation levels. Increasing the spacing from 5 mm +5 mm to 7 mm +7 mm causes higher cross-polarized radiation levels and a narrower impedance bandwidth due to the larger inductance. The cross-polarized radiation levels decrease but the impedance bandwidth is reduced when the spacing is decreased from 5 mm +5 mm to 4.5 mm +4.5 mm.

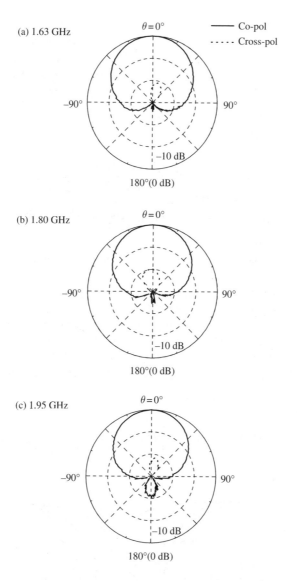

Figure 3.41 Measured co- and cross-polarized radiation patterns in the H-plane. (Reproduced by permission of IEEE.[40])

3.3.5 SPA WITH PROBE-FED CENTER SLOT

As discussed in earlier sections, the dual-probe and half-wavelength probe-fed strip schemes can be employed to improve the radiation performance of broadband SPAs because of the balanced or balance-like structures. On the other hand, as mentioned in section 3.3.2, the dual-probe scheme suffers from two drawbacks. One is that two feed probes are needed; the other is that in the cuts off the E- and H-planes, the cross-polarized radiation may not be cancelled out completely. Furthermore, as can be seen from section 3.3.4, an SPA driven by

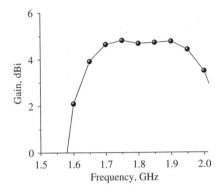

Figure 3.42 Measured average gain of the SPA fed by a half-wavelength strip.

a half-wavelength probe-fed strip distorts the co-polarized radiation patterns in the E-plane owing to the asymmetrical structure.

A feeding technique with a probe-fed center slot is presented now to alleviate these problems. Basically, the aim of the center-fed (CF) feeding scheme is to achieve a balance-like feeding structure by exciting a radiating plate centrally and symmetrically. The CF-SPAs are expected to improve radiation performance by having a high co-to-cross-polarized ratio and acceptable co-polarized radiation patterns across a broad bandwidth. Also, the feeding structure should be simple and able to achieve impedance matching easily.

CF-SPA with Shorting Strip

Figure 3.43 shows the rectangular CF-SPA, which comprises a perfectly electrically conducting (PEC) plate of length L and width $W = 70$ mm parallel to the ground plane at a

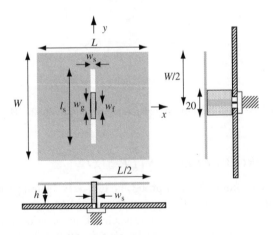

Figure 3.43 Geometry of a center-fed SPA (dimensions in millimeters). (Reproduced by permission of IEEE.[41])

height $h = 9$ mm. A rectangular PEC plate measuring 305 mm × 225 mm is used. The plate will be centrally fed to keep the overall configuration symmetrical. However, it is difficult to excite the dominant TM_{01} mode when the feed point is located at the midpoint of the radiating plate. Therefore, a narrow rectangular slot with dimensions $l_s \times w_s$ (1 mm ×50 mm) is cut centrally from the plate. The longer sides of the slot are parallel to the radiating edges of the plate.

The feeding structure is made of two parallel strips etched back-to-back on a thin dielectric slab measuring 20 mm ×10 mm ×32 mil (about 0.81mm) (Roger 4003, $\epsilon_r = 3.38$). Both the strips are electrically connected to the longer sides of the slot at their upper ends. The bottom of the strips are, respectively, grounded and driven by a 50-Ω coaxial probe through an SMA connector. The widths of the probe-driven and grounded strips are w_f and w_g, respectively. The strips mainly operate in two modes, namely a non-radiating transmission-line mode with equal magnitudes but opposite phase, and a radiating antenna mode with unequal magnitudes but in phase. By changing the width ratio (w_g/w_f) and the spacing between the strips, the operating modes can be controlled. A larger width ratio results in weaker radiation from the strips in order to form a balance-like feeding structure as it operates almost in the non-radiating mode. To keep the induced electric currents on the plate symmetrical, a tradeoff between forming an electrically balanced and structurally symmetrical feed was made by selecting the ratios 2:1 and 1:1.

Figure 3.44 shows the measured VSWR of three CF-SPAs with different lengths L and width ratios w_g/w_f. It is seen that the three designs have nearly the same 2:1 VSWR bandwidths of about 9.5 %. The resonant frequencies of Antennas 1, 2 and 3 are 1.72, 1.64 and 1.69 GHz, which correspond to the frequencies at which the measured VSWR is minimum. The length L of Antenna 2 is set to be 75 mm to have its resonant frequency nearly the same as the others. This also suggests that the feeding structure operates in a transmission-line mode rather than a monopole mode owing to the reduced effective radiating length of the antenna.

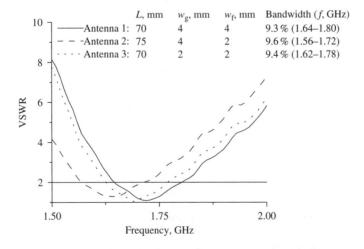

	L, mm	w_g, mm	w_f, mm	Bandwidth (f, GHz)
——Antenna 1:	70	4	4	9.3 % (1.64–1.80)
– – -Antenna 2:	75	4	2	9.6 % (1.56–1.72)
· · · · Antenna 3:	70	2	2	9.4 % (1.62–1.78)

Figure 3.44 Measured VSWR of the three center-fed SPAs. (Reproduced by permission of IEEE.[41])

Figure 3.45 and 3.46 show the radiation patterns of Antenna 1 at 1.64, 1.72 and 1.80 GHz. By collating the co- and cross-polarized radiation patterns in the E-plane (x–z plane) as shown in Figure 3.45, it can be found that the co-to-cross-pol ratios are greater than 20 dB in all directions, and the shapes of the co-polarized radiation patterns are hardly changed with 70° half-power beamwidths. Similarly, Figure 3.46 shows the co- and cross-polarized radiation patterns in the H-plane (y–z plane). At all the frequencies, the shapes of both co- and cross-polarized radiation patterns are unchanged. The half-power beamwidths of the co-polarized radiation patterns are about 60° at all the three frequencies. The maximum cross-polarized radiation appears at around the $\theta = \pm 40°$ directions, and the co-to-cross-pol ratios are more than 20 dB.

The measured results demonstrate that the CF feeding technique can be used to mitigate the degradation of radiation performance of SPAs in two ways. One is to suppress the cross-polarized radiation by making the cross-polarized currents symmetrical with respect to the midlines of the radiating plate but out of phase. The other method is to improve the co-polarized radiation patterns by making the co-polarized currents symmetrical about the

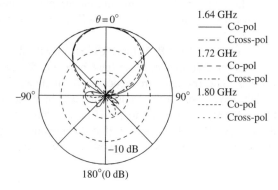

Figure 3.45 Measured co- and cross-polarized radiation patterns in the E-plane. (Reproduced by permission of IEEE.[41])

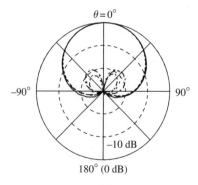

Figure 3.46 Measured co- and cross-polarized radiation patterns in the H-plane. (Reproduced by permission of IEEE.[41])

midlines of the plate. Compared with a dual-probe feeding structure, this feeding structure located at the center of the plate features a simple feed design with only one balance-like feed and no additional feeding networks. As the feed point is located at the center of the radiating plate, the cross-polarized currents due to the higher-order modes are around the center slot, and the cross-polarized components at the two sides of the center slot are out of phase. As a result, the cross-polarized radiation is significantly suppressed in all ϕ-cuts.

Furthermore, according to the equivalence theorem, the probe-fed slot at the plate can be considered as an equivalent magnetic current source, which excites the plate centrally and symmetrically. The feeding structure is similar to the aperture coupling structure, where the equivalent magnetic current source is located at the ground plane.

The effects of varying the geometric parameters of the CF-SPA on the impedance characteristics are considered next. In the study, the plate is fed by a pair of cylindrical wires with radii 0.5 mm instead of the strips, and the geometric parameters of the antenna are set to be $W = L = 70$ mm, $h = 9$ mm, and $l_s \times w_s = 1$ mm $\times 50$ mm. The frequency range is from $f_{start} = 1.5$ GHz to $f_{stop} = 2.0$ GHz.

Figure 3.47 shows the effects of varying spacing h from 7 mm to 12 mm on the input impedance. The impedance loci demonstrate that the input inductance greatly increases when the spacing becomes larger. The impedance loci for the different slot lengths, $l_s = 40$ mm to 60 mm, are shown in Figure 3.48. Evidently, the input impedance is quite sensitive to l_s. A long slot leads to an inductive input impedance. In contrast to the aperture-coupled patch antenna, the size of the slot on the plate affects not only the impedance matching but also the resonant frequency of the antenna. Figure 3.49 shows the impedance loci as the slot width w_s is varied from 1 mm to 3 mm. Clearly, the input impedance is not sensitive to w_s. This suggests that good impedance matching can be obtained across a broad bandwidth by adjusting the spacing h or/and the length l_s. In particular, by increasing the slot length, the large inductance due to the large spacing can be compensated, and good impedance matching can be achieved.

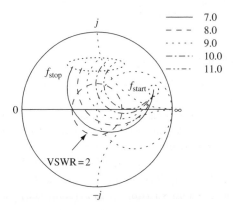

Figure 3.47 Effects of varying h on the impedance performance of the CF-SPA (dimensions in millimeters). (Reproduced by permission of IEEE.[41])

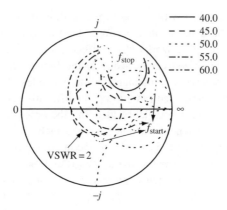

Figure 3.48 Effects of varying l_s on the impedance performance of the CF-SPA (dimensions in millimeters). (Reproduced by permission of IEEE.[41])

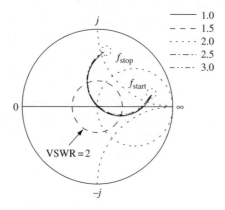

Figure 3.49 Effects of varying w_s on the impedance performance of the CF-SPA (dimensions in millimeters). (Reproduced by permission of IEEE.[41])

CF-SPA with Resistive/Capacitive Loading

The short-circuited strip in the CF-SPA can be replaced by a resistive or capacitive loading. The impedance and radiation characteristics of these CF-SPAs, such as impedance bandwidth, gain, radiation patterns and cross-polarization radiation levels, will be discussed. In this study, the CF-SPAs are fed by a single probe, short-circuited and open-circuited loading, as well as resistive and capacitive loading, respectively.[42]

Figure 3.50 shows the geometry of a probe-fed CF-SPA suitable for use with a chip resistor or chip capacitor loading. The chip load is soldered across a 1-mm gap on a 4-mm wide conducting strip etched on a 10 mm × 10 mm × 32 mil (about 0.81mm) square dielectric sheet (Roger4003, $\epsilon_r = 3.38$). The strip is grounded at one end, and connected to the radiating patch at the other. The radiating plate is centrally fed at the edge of the 1-mm center slot of length l_s by a 50-Ω SMA via a 2-mm wide strip that runs on the other side of the dielectric

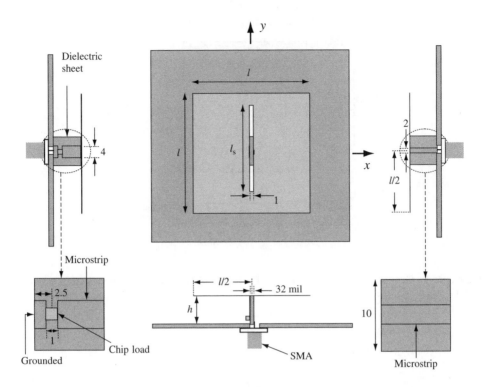

Figure 3.50 Geometry of the probe-fed CF-SPA with chip loading (dimensions in millimeters). (Reproduced by permission of John Wiley & Sons, Inc.[42])

sheet. The square radiating plate positioned in the x–y plane has a length $l = 70$ mm and is suspended at a height of $h = 10$ mm.

The geometry of an SPA loaded by a perfectly electrically conducting (PEC) plate with dimensions $l_{cap} \times w_{cap}$ at a height h_{cap} above the ground plane is shown in Figure 3.51. Supported with a foam layer, this plate capacitor is electrically connected at one edge to the trimmed end of the 4-mm wide strip in place of the chip capacitor used in Figure 3.50. A PEC plate with dimensions 305 mm $\times$ 305 mm is used to approximate the infinite ground plane in the measurements.

For comparison, three other CF-SPAs with no shorting pin (only a feeding probe), short-circuited loading, and open-circuited loading are studied as well. By varying the load resistance, the impedance bandwidth is optimized by adjusting the length of the center slot. As a result, the impedance bandwidth is maximum when the load is 39 Ω. Figure 3.52 compares the measured return loss of the SPA with the 39-Ω load (Antenna 4) and the other SPAs (Antennas 1–3).

The capacitive loading can be implemented by using either a plate capacitor or a chip capacitor. A plate capacitor (Figure 3.51) is used to broaden the impedance bandwidth. The optimal dimensions of the plate capacitor are $l_{cap}(= w_{cap}) = 16$ mm and $h_{cap} = 1$ mm for Antenna 5, and $l_{cap}(= w_{cap}) = 38$ mm and $h_{cap} = 5$ mm for Antenna 6, yielding capacitances of 2.27 pF and 2.56 pF, respectively. The other capacitive-loaded antenna is implemented using a 2.2-pF chip capacitor (Antenna 7) as shown in Figure 3.50. The chip capacitor features

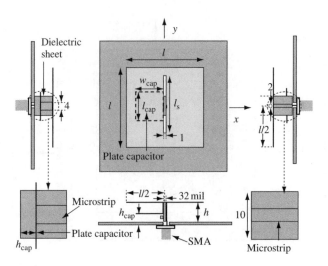

Figure 3.51 Geometry of the probe-fed CF-SPA with a plate capacitor. (Reproduced by permission of John Wiley & Sons, Inc.[42])

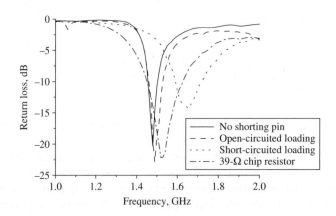

Figure 3.52 Comparison of measured return loss of the probe-fed CF-SPA with the 39-Ω load (Antenna 4) and three other SPAs (Antennas 1–3). (Reproduced by permission of John Wiley & Sons, Inc.[42])

small size and easy installation. The impedance bandwidth of Antenna 7 increases slightly to 7.4 % with a slight decrease in resonant frequency. Figure 3.53 shows the measured return losses for Antennas 2 and 5–7.

By adjusting the length of the center slot l_s, good impedance matching can be obtained. Table 3.3 compares the measured impedance bandwidths and resonant frequencies. Among them, Antenna 4 with a resistive load is capable of achieving the largest impedance bandwidth of more than 10 % for a return loss of less than −10 dB, owing to the reduction in the Q of

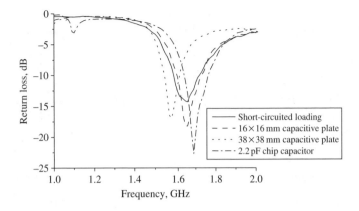

Figure 3.53 Geometry of the probe-fed CF-SPA with a plate capacitor. (Reproduced by permission of John Wiley & Sons, Inc.[42])

Table 3.3 Measured impedance bandwidths (BW) at the lower (f_1), center (f_c) and upper (f_u) frequencies, as well as the gain at f_c for CF-SPAs with various loadings.

Antenna	Loading	l_s (mm)	f_1 (GHz)	f_c (GHz)	f_u (GHz)	BW (%)	Gain (dBi)
1	No shorting pin	62	1.46	1.48	1.50	2.7	7.3
2	Short-circuited	58	1.60	1.66	1.71	6.7	7.3
3	Open-circuited	60	1.39	1.41	1.44	3.5	6.9
4	$R_{load} = 39\ \Omega$	68	1.46	1.52	1.62	10.4	4.6
5	$l_{cap} \times w_{cap}=$ 16 mm ×16 mm	60	1.61	1.66	1.72	6.6	7.7
6	$l_{cap} \times w_{cap}=$ 38 mm ×38 mm	60	1.55	1.58	1.62	4.4	7.7
7	C_{load}=2.2 pF	62	1.64	1.69	1.76	7.4	6.7

the antenna. Antennas 2, 5 and 7 with a shorting pin and capacitive loading have attained more than 6 % bandwidths.

The simulated and measured co- and cross-polarized radiation patterns for Antennas 1–7 at the lower (f_1), center (f_c) and upper (f_u) frequencies of the bandwidth are illustrated in Figures 3.54–3.60. The measured cross-polarized radiation levels are slightly higher than the simulated ones owing to the finite ground plane and the RF signal cable in the tests. The figures show that, in the H-plane, the co-polarized radiation patterns for all the SPAs are symmetrical with respect to the boresight. In the E-plane, the co-polarized radiation patterns, except for those for Antenna 6, are found to be slightly asymmetric owing to the asymmetrical feed configuration with respect to the H-plane. The cross-polarized radiation patterns in both the E- and H-planes have the same features as those for conventional microstrip patch antennas.

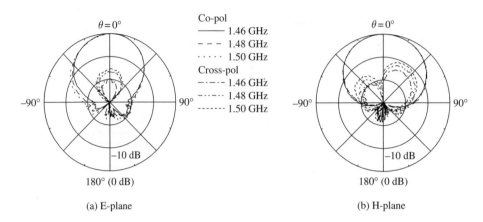

Figure 3.54 Measured radiation patterns for Antenna 1 (without a shorting pin) at 1.46, 1.48 and 1.50 GHz. (Reproduced by permission of John Wiley & Sons, Inc.[42])

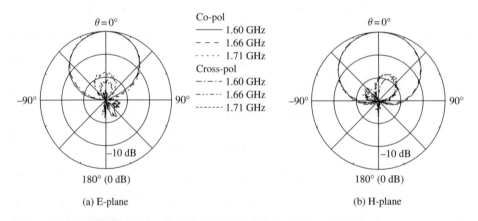

Figure 3.55 Measured radiation patterns for Antenna 2 (with a short-circuited load) at 1.60, 1.66 and 1.71 GHz. (Reproduced by permission of John Wiley & Sons, Inc.[42])

The measured gain for the SPAs with various loadings is given in Table 3.3. In contrast to the gain of 7.3 dBi for Antenna 2, Antennas 5 and 6 with plate capacitors have higher gain of 7.7 dBi. Antenna 4 with a bandwidth of 10.4 % has the lowest gain of 4.4 dBi owing to the resistive load.

As shown in Table 3.4, the co-to-cross-pol ratios in the E-plane are higher than those in the H-plane. The co-to-cross-pol ratios of 23.4/19.4 dB (in the E/H-planes) at the center frequency for Antenna 6 are significantly higher than for the others. The half-power beamwidths of the SPAs are about 58° in the E-plane and 70° in the H-plane. Therefore, Antenna 6 is the best option in terms of the gain and the co-to-cross-pol ratio, whereas Antenna 4 has the broadest impedance bandwidth but the lowest gain. Antenna 2 has a

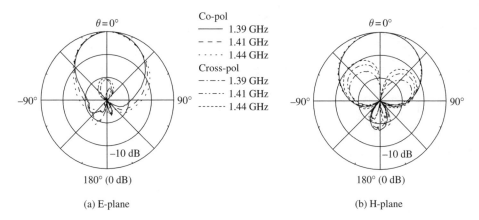

Figure 3.56 Measured radiation patterns for Antenna 3 (with an open-circuited load) at 1.39, 1.41 and 1.44 GHz. (Reproduced by permission of John Wiley & Sons, Inc.[42])

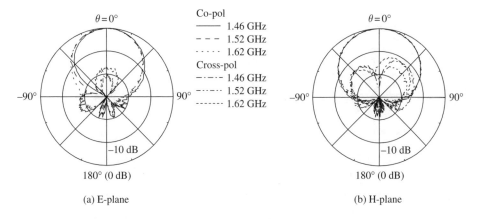

Figure 3.57 Measured radiation patterns for Antenna 4 (with a 39-Ω chip resistor load) at 1.46, 1.52 and 1.62 GHz. (Reproduced by permission of John Wiley & Sons, Inc.[42])

moderate performance in terms of impedance bandwidth, and co-to-cross-pol ratios with reasonable gain.

This study of the characteristics of CF-SPAs with resistive and capacitive loads has shown that a CF-SPA with an optimized resistive load is capable of providing a broad impedance bandwidth but with a low gain and low co-to-cross-pol ratios across the bandwidth. The optimized CF-SPA with a plate capacitor load is capable of increasing the gain and co-to-cross-pol ratios but results in a low impedance bandwidth. In terms of impedance bandwidth, gain, co-to-cross-pol ratio as well as design complexity, a CF-SPA with a short-circuited load is considered to be the best option.

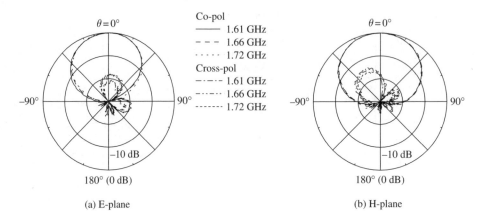

Figure 3.58 Measured radiation patterns for Antenna 5 (with a 16 mm ×16 mm plate capacitor load) at 1.61, 1.66 and 1.72 GHz. (Reproduced by permission of John Wiley & Sons, Inc.[42])

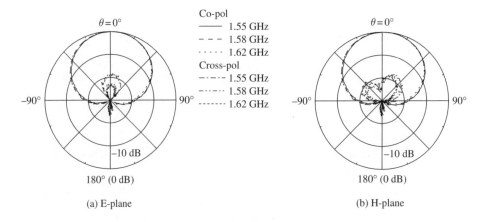

Figure 3.59 Measured radiation patterns for Antenna 6 (with a 38 mm ×38 mm plate capacitor load) at 1.55, 1.58, and 1.62 GHz. (Reproduced by permission of John Wiley & Sons, Inc.[42])

3.3.6 CASE STUDY: CENTER-FED SPA WITH DOUBLE L-SHAPED PROBES

The use of dual out-of-phase probes allows an improvement in radiation performance because of the cancellation of the cross-polarized radiation. However, a broadband feeding network must be used to keep the two probes out-of-phase across the bandwidth. For single-element applications, this increases the design complexity. Therefore, an SPA fed by a pair of L-shaped strips is used, which combines the advantages of the symmetrical balanced center-fed design[41] and the use of an L-shaped feed probe.[21] The symmetrical, balanced and center-fed design is capable of providing acceptable radiation performance by mitigating the

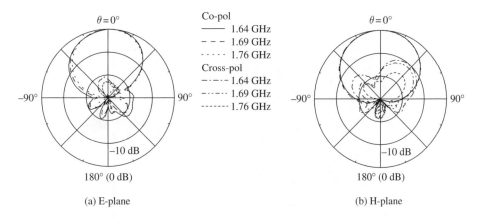

(a) E-plane   (b) H-plane

Figure 3.60 Measured co- and cross-polarization radiation patterns for Antenna 7 (with a 2.2-pF chip capacitor load) at 1.64, 1.69, and 1.76 GHz. (Reproduced by permission of John Wiley & Sons, Inc.[42])

Table 3.4 Measured co-to-cross-pol ratios at the lower (f_1), center (f_c), and upper (f_u) frequencies for CF-SPAs with various loadings.

Antenna	Loading	Co-to-cross-pol ratio in E-planes (dB)			Co-to-cross-pol ratio in H-planes (dB)		
		At f_1	At f_c	At f_u	At f_1	At f_c	At f_u
1	No shorting pin	18.9	16.6	18.3	11.2	9.4	7.9
2	Short-circuited	18.9	16.6	18.3	11.2	9.4	7.9
3	Open-circuited	18.9	16.6	18.3	11.2	9.4	7.9
4	$R_{load} = 39\,\Omega$	18.9	16.6	18.3	11.2	9.4	7.9
5	$l_{cap} \times w_{cap} =$ 16 mm × 16 mm	18.9	16.6	18.3	11.2	9.4	7.9
6	$l_{cap} \times w_{cap} =$ 38 mm × 38 mm	18.9	16.6	18.3	11.2	9.4	7.9
7	$C_{load} = 2.2\,pF$	18.9	16.6	18.3	11.2	9.4	7.9

cross-polarized radiation and improving the co-polarized radiation patterns. The L-shaped probes allow a broad impedance bandwidth through the electromagnetic coupling between the probes and the radiator.

Figure 3.61 shows the geometry.[43] The coordinate system is oriented so that the radiating square plate lies in the x–y plane and its z-axis is vertically centered through the structure. The 70 mm ×70 mm plate is placed above a 310 mm ×310 mm ground plane at a height H. A 1 mm×50-mm rectangular slot is cut centrally from the plate with the longer sides parallel to the y-axis. The distance between the slot's midline and the longer side along y-axis is denoted by s. The center slot enhances the radiation performance at higher operating

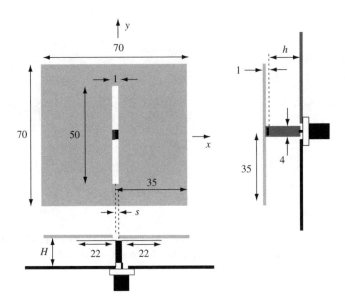

Figure 3.61 Geometry of an SPA fed by a pair of L-shaped probes (dimensions in millimeters). (Reproduced by permission of IEEE.[43])

frequencies by modifying the distribution of the electric currents induced on the plate.[41] By adjusting the length of the slot, good impedance matching can be achieved.

The feeding structure consists of a pair of 4-mm wide L-shaped strips (similar to a horizontal center-fed dipole), standing back-to-back and centrally installed beneath the plate. Thus, the antenna configuration is symmetrical with respect to both x- and y-axes as $s = 0$ mm, which can improve the co-polarized patterns and the suppress the cross-polarized radiation. Each of the L-shaped strips comprises a horizontal and a vertical arm of height h. One of the vertical arm is grounded and the other fed by a 50-Ω coaxial probe at the bottom. The two vertical arms are separated by a piece of 1.5-mm thick and 4-mm wide Rogers4003 PCB ($\epsilon_r = 3.38$). The PCB piece has the same height h as the vertical arm. The lengths of the two horizontal arms are 22 mm, which have been numerically optimized for the height $H = 10$ mm, and placed along the x-axis. The grounded and probe-excited strips form a balance-like feeding structure, carrying the out-of-phase electric currents with almost the same amplitude. The gap between the plate and the horizontal arms is set to be 1 mm for the L-shaped strips to have a strong electromagnetic coupling with the radiating plate so as to broaden the impedance bandwidth.[21] It has been observed from both simulations and experiments that the lengths of the horizontal arms and the gap have a significant effect on impedance matching.

The impedance and radiation performance has been examined for heights H of 10, 15, and 20 mm. The measured and simulated results in Table 3.5 show that the distance s affects the impedance matching condition significantly. For brevity, Figure 3.62 displays only the measured and simulated VSWR of the SPAs with $s = 2$ mm across the frequency range 0.5–2.5 GHz. The figure shows two well-matched bands for each of the heights H.

Table 3.5 Measured impedance bandwidths for SPAs with varying heights.

Antenna	H(mm)	s(mm)	Bandwidth (VSWR = 2:1)	Test frequencies(GHz)
1	10	2	17 % (1.65–1.97 GHz)	1.7, 1.8, 1.9, 2.0
2	15	3	26 % (1.46–1.87 GHz)	1.5, 1.6, 1.7, 1.8, 1.9
3	20	4	25 % (1.44–1.85 GHz)	1.5, 1.6, 1.7, 1.8, 1.9

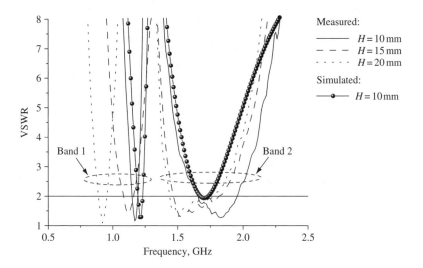

Figure 3.62 Measured VSWR of an SPA fed by a pair of L-shaped probes, with $s = 2$. (Reproduced by permission of IEEE.[43])

The measured VSWR is plotted against the distance s for varying heights H in Figure 3.63. The 2:1 VSWR bandwidth for Band 1 are 3–6 % but reaches up to 17 % for $H = 10$ mm and to 25 % for $H = 15$ mm or 20 mm for Band 2. In Band 2, the achieved bandwidth with $H = 10$ mm and 15 mm are sensitive to the distance s and the widest bandwidth are attained when 0 mm $\leq s \leq 4$ mm. This may be because the two feeding strips are not completely electrically balanced. However, the 2:1 VSWR bandwidth for $H = 20$ mm is always more than 22 % for -6 mm $\leq s \leq 4$ mm. In addition, the electromagnetic coupling by the use of the L-shaped probes evidently broadens the impedance bandwidth.

Figure 3.64 displays the measured and simulated radiation patterns for the SPA with $s = 2$ mm and $H = 10$ mm in the x–z and y–z planes. The simulated E_ϕ components in the E-plane are much lower that the measured ones. The radiation patterns for the E_θ components in both the E- and H-planes indicate that the radiation characteristics of the SPA are quite similar to those of a vertical T-shaped monopole.

Next, the radiation patterns for the three designs are measured in two principal E- and H-planes at specific frequencies within Band 2 as listed in Table 3.5. Only the radiation patterns at the highest frequencies within Band 2 are included because in the lower band

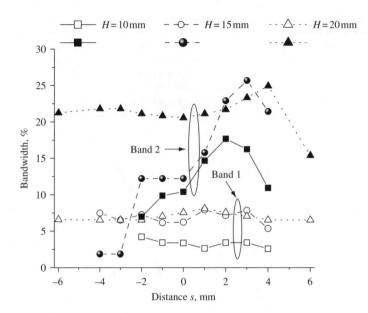

Figure 3.63 Measured impedance bandwidths for the SPAs with heights $H = 10$, 15, and 20 mm. (Reproduced by permission of IEEE.[43])

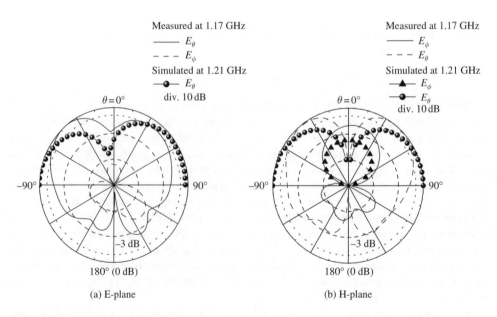

Figure 3.64 Measured co- and cross-polarized radiation patterns in the E- and H-planes for H=10 mm and S=2 mm. (Reproduced by permission of IEEE.[43])

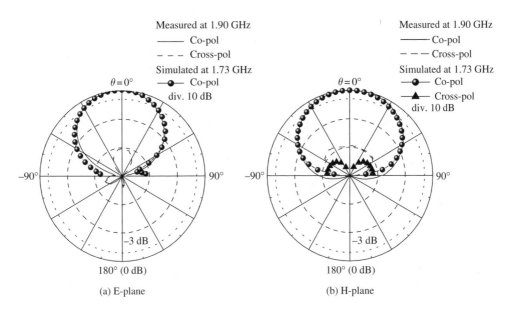

(a) E-plane

(b) H-plane

Figure 3.65 Radiation patterns in the E- and H-planes for $H = 10$ mm and $s = 2$ mm. (Reproduced by permission of IEEE.[43])

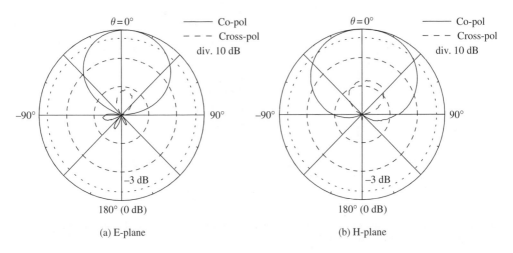

(a) E-plane

(b) H-plane

Figure 3.66 Measured radiation patterns in the E- and H-planes for $H = 15$ mm and $s = 3$ mm at 1.80 GHz. (Reproduced by permission of IEEE.[43])

(Band 1), the SPAs act as top-loaded monopoles, and in the upper band (Band 2) they operate in the dominant modes of patch antennas with electrically larger sizes. The co- and cross-polarized radiation patterns are illustrated in Figures 3.65–3.67. Other radiation factors, such as gain, half-power beamwidth and the co-to-cross-pol ratio across the bandwidth are plotted in Figures 3.68–3.70.

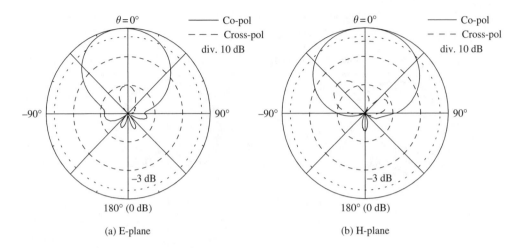

(a) E-plane   (b) H-plane

Figure 3.67 Measured radiation patterns in the E- and H-planes for $H = 20$ mm, $s = 3$ mm at 1.80 GHz. (Reproduced by permission of IEEE.[43])

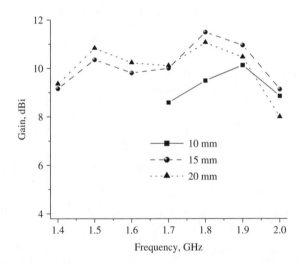

Figure 3.68 Comparison between the measured gain of the SPAs with varying heights H. (Reproduced by permission of IEEE.[43])

Figure 3.65 shows the measured and simulated radiation patterns in the E- and H-planes for $H = 10$ mm. The simulated cross-polarized components are much lower than the measured ones, especially in the E-planes. Figures 3.66 and 3.67 show, respectively, the measured radiation patterns for the SPAs with $H = 15$ mm and 20 mm, and $s = 3$ mm. It can be seen from Figures 3.65–3.67 that, over Band 2, the co-polarized radiation patterns in the E- and H-planes are almost unchanged for varying H, having symmetrical shapes with respect to the boresight and without additional dips. The side lodes and back lobes are much lower

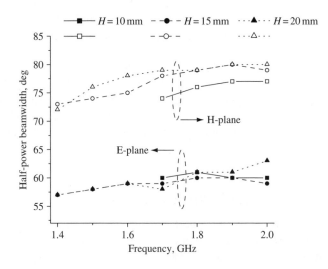

Figure 3.69 Comparison between the measured beamwidths of the SPAs fed at varying heights H. (Reproduced by permission of IEEE.[43])

than -25 dB. However, the measured cross-polarized radiation levels in Figures 3.66 and 3.67 have a 1–2 dB error at the boresight owing to the received power levels being too low.

Figure 3.68 demonstrates the average gain of about 9 dBi for $H = 10$ mm and about 10 dBi for $H = 15$ mm and 20 mm. Within the bandwidth, the gain of the SPA is greater than 8.5 dBi for $H = 10$ mm. For $H = 15$ mm and 20 mm, the gain is greater than 9.2 dBi. The half-power beamwidth range from 70° to 80° in the H-plane and from 55° to 65° in the E-plane, as illustrated in Figure 3.69.

Figure 3.70 shows that the co-to-cross-pol ratios are more than 20 dB in the E-plane and vary between 16 dB and 21 dB in the H-plane. Compared with conventional SPAs,

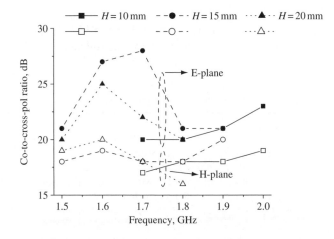

Figure 3.70 Comparison between measured co-to-cross-pol ratios of the SPAs fed at varying heights H. (Reproduced by permission of IEEE.[43])

the radiation performance of these SPAs is considerably improved across the bandwidth of 20 %. In principle, the enhancement of the radiation performance stems mainly from the symmetrical balanced center-fed configuration.

However, it is important to note that the cross-polarized radiation caused by the higher-order modes at the higher frequencies is not completely cancelled out by the balance-like feeding design suggested to replace the ideal balanced feeding structure. The simplified balance-like design results in a slightly asymmetrical configuration with the distance $s \geq 0$ mm in order to offset the asymmetry of the electric currents on the horizontal arms of the two L-shaped strips.

3.3.7 SPA WITH SLOTS AND SHORTING STRIPS

SPAs with thick and low-permittivity substrates feature broad impedance bandwidths but degraded radiation performance. The earlier discussions have shown that the careful selection of the feeding scheme and the use of loadings can effectively enhance the impedance and radiation performance. Usually, the use of shorting pins or slots is a simple but effective way.

Shorting pins are often used to reduce the size or improve the impedance matching of microstrip antennas.[44] Adding the shorting pins, often in the form of a strip, is also expected to improve the radiation performance of a strip-fed SPA instead of using out-of-phase feed strips in the dual-probe scheme. The shorting strip grounds the SPA at the location opposite to a feed point to form a symmetrical structure. Besides the use of shorting pins, slotting radiators is one of the important techniques to realize impedance matching, reduce the size, and achieve broad or multiple band operation in microstrip antennas.[45,46] The hybrid use of shorting strips and narrow slots for enhancing the impedance and radiation performance of SPAs will be discussed in this section.

Description of the Antennas

Table 3.6 lists the eight SPAs under consideration, and Figure 3.71 shows their configurations and the coordinate system. Antennas 2–7 are variations of Antennas 0 or 1. A square copper

Table 3.6 Description of SPAs with shorting pins and/or slots.

Antenna	Description
0	Strip-fed without any shorting strip
1	Strip-fed with a shorting strip
2	Strip-fed with a center slot but without a shorting strip
3	Strip-fed with a center slot and a shorting strip
4	Strip-fed with two edge slots but without a shorting strip
5	Strip-fed with two edge slots and a shorting strip
6	Strip-fed with two edge slots and a center slot without a shorting strip
7	Strip-fed with two edge slots, a center slot and a shorting strip

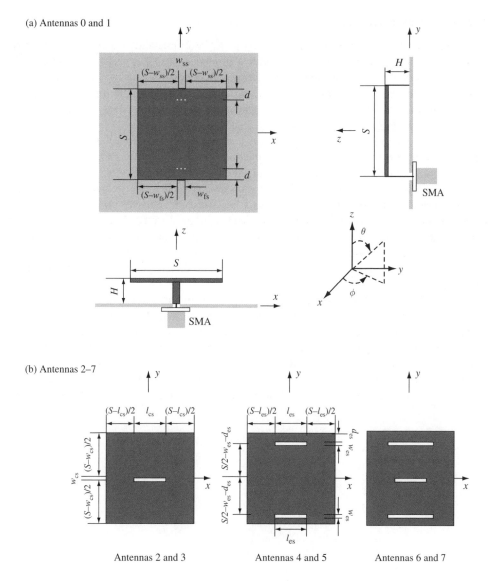

Figure 3.71 Geometry of the SPAs with shorting strips and/or slots. (Reproduced by permission of © IEEE. Z. N. Chen, 'Suspended plate antennas with shorting strips and slots,' *IEEE Transactions on Antennas and Propagation*, vol. 52, no. 10, pp. 2525–2531, October 2004.)

plate ($S \times S = 70\,\text{mm} \times 70\,\text{mm}$) is laid parallel to a ground plane (305 mm ×305 mm) at a height H. Usually, a square plate antenna generates higher cross-polarized radiation and is suitable for assessing techniques to suppress the cross-polarized radiation levels. A 2-mm wide probe-fed strip is vertically positioned and excites the center of one of the plate edges. Similarly, another 2-mm wide strip grounds the center of the opposite edge of the plate. As such, the feed and shorting strip form a structurally symmetrical SPA.

Impedance Characteristics

Antennas 0 and 1. Antenna 0 is fed by a strip ($w_{fs} = 2\,mm$) and is a conventional SPA. Antenna 1 is a variation of Antenna 0 with a grounded strip ($w_{ss} = 2\,mm$). Figure 3.72 shows the simulated and measured VSWR against frequency, where the height H is 8 mm and the distance d is 0 mm. The measured and simulated bandwidths for VSWR $= 2$ and resonant frequencies (at which the VSWR reaches minima) are, respectively, 7.2 %/6.5 % and 1.92 GHz/1.96 GHz for Antenna 0, and 5.9 %/6.3 % and 2.12 GHz/2.11 GHz for Antenna 1.

Introducing a shorting strip slightly reduces the bandwidth. It is important to note that the narrow shorting strip does not result in a reduction in the resonant frequency of the SPA by half.[44] Figure 3.73 shows the distribution of the simulated imaginary parts of the induced electric currents on the plates since the radiation is attributed to the imaginary parts of the induced currents. Unlike Antenna 0, Antenna 1 has x-directed currents concentrated on each of the radiating edges where the shorting and feed strips are located. The x-directed currents close to the radiating edges of Antenna 1 are out of phase with respect to the y-axis, so that the radiation from the x-directed currents will be cancelled out in the E-plane. This distortion reduces the effective radiating length of Antenna 1, and thus slightly increases the resonant frequency.

Figure 3.74 shows the effects of varying the width of the shorting strip on impedance matching when the feed strip is located at the edge of the plate ($d = 0\,mm$). Increasing the width w_{ss} shifts up the resonant frequency and a resonance is excited at around 1 GHz. By moving the feed strip and the shorting strip closer ($d > 0\,mm$) to each other, impedance matching can be achieved at lower frequencies. This suggests that the narrow shorting strip and the shorting wall used in [44] play distinct roles in these SPAs. The former acts as an impedance tuner in a planar inverted-F antenna operating in the monopole-like mode since

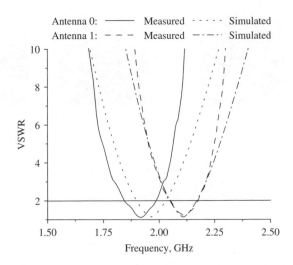

Figure 3.72 Comparison between the measured and simulated VSWR for Antennas 0 and 1. (Reproduced by permission of © IEEE. Z. N. Chen, 'Suspended plate antennas with shorting strips and slots,' *IEEE Transactions on Antennas and Propagation*, vol. 52, no. 10, pp. 2525–2531, October 2004.)

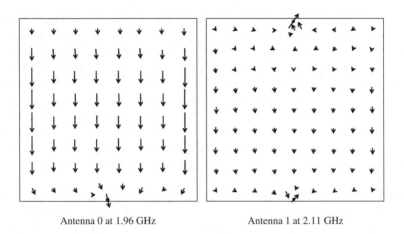

Antenna 0 at 1.96 GHz   Antenna 1 at 2.11 GHz

Figure 3.73 Simulated distributions of the imaginary parts of electric currents for Antennas 0 and 1 operating at resonant frequencies. (Reproduced by permission of © IEEE. Z. N. Chen, 'Suspended plate antennas with shorting strips and slots,' *IEEE Transactions on Antennas and Propagation*, vol. 52, no. 10, pp. 2525–2531, October 2004.)

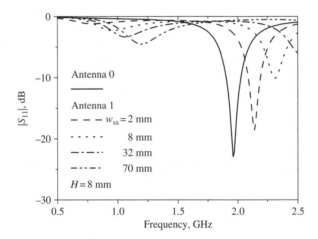

Figure 3.74 Simulated return losses against frequency for Antennas 0 and 1 with various strip widths. (Reproduced by permission of © IEEE. Z. N. Chen, 'Suspended plate antennas with shorting strips and slots,' *IEEE Transactions on Antennas and Propagation*, vol. 52, no. 10, pp. 2525–2531, October 2004.)

the feed probe is rather close to the shorting wall. The latter acts as a load of the microstrip antenna operating in a dominant cavity mode, which will be verified by the measured radiation patterns later.

Antennas 2 and 3. As known, the introduction of a center slot can lower the resonant frequency of an SPA. Antenna 2 is a variation of Antenna 0, with a narrow center slot ($w_{cs} = 2$ mm, $l_{cs} = 0 - 68$ mm). Antenna 3 is a variation of Antenna 2, grounded by a shorting strip ($w_{ss} = 2$ mm) as shown in Figure 3.71.

Figure 3.75 shows the measured and simulated return losses versus frequency for Antennas 2 and 3. The simulations show that for $l_{cs} < 35$ mm, the resonant frequencies are virtually unchanged around 2.1 GHz and the impedance matching remains very good. However, the resonant frequencies are reduced by more than 10 % (from 1.99 GHz to 1.78 GHz) as the slot length increases from 40 mm to 68 mm. A slot length of 40 mm is selected for subsequent analysis. The measured and simulated bandwidths $|S_{11}| < -10$ dB and resonant frequencies are, respectively, 3.0 %/2.9 % and 1.82 GHz/1.81 GHz for Antenna 2, and 5.2 %/4.9 % and 2.01 GHz/1.99 GHz for Antenna 3. Compared with Antenna 1, the center slot at Antenna 3 leads to a 5 % (about 110 MHz) decrease in the resonant frequency, which can be used to compensate for the increase in the resonant frequency due to the shorting strip.

Antennas 4 and 5. Cutting a pair of narrow rectangular edge slots in a planar radiator has been used to realize a dual-band design.[45,46] Antenna 4 is a variation of Antenna 0 with two slots of the same lengths l_{es} and widths $w_{es} = 2$ mm. The slots are positioned 4 mm from the edges with the longer sides parallel to the edges. Antenna 4 grounded by a shorting strip forms Antenna 5. The height H is kept at 8 mm. The effects of varying the length of the edge slots ranging from 20 mm to 60 mm on the impedance characteristics of Antennas 4 and 5 are plotted in Figure 3.76. Within the range of 1 GHz to 3 GHz, Antennas 4 and 5 with $l_{es} = 60$ mm have two resonances because of the edge slots.[45,46] The measured and simulated lower resonant frequencies of Antenna 5 are 2.15 GHz and 2.13 GHz for the optimal slot length of 60 mm. The simulation shows that the edge slot hardly affects the lower resonant frequency, and increasing the length of the edge slots greatly reduces the upper resonant frequency. The measured and simulated impedance bandwidths of the dominant band for $|S_{11}| < -10$ dB are, respectively, 3.9 %/3.4 % and 3.9 %/4.7 % for Antennas 4 and 5, which are slightly narrower than those for Antennas 0 and 1.

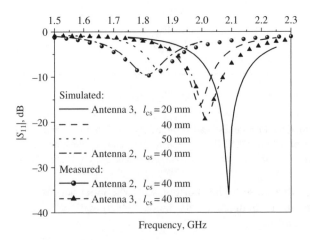

Figure 3.75 Measured and simulated return losses against frequency for Antennas 2 and 3 with various center slot lengths. (Reproduced by permission of © IEEE. Z. N. Chen, 'Suspended plate antennas with shorting strips and slots,' *IEEE Transactions on Antennas and Propagation*, vol. 52, no. 10, pp. 2525–2531, October 2004.)

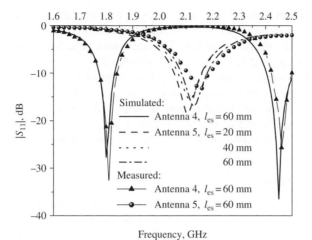

Figure 3.76 Measured and simulated return losses against frequency for Antennas 4 and 5 with various edge slot lengths. (Reproduced by permission of © IEEE. Z. N. Chen, 'Suspended plate antennas with shorting strips and slots,' *IEEE Transactions on Antennas and Propagation*, vol. 52, no. 10, pp. 2525–2531, October 2004.)

Antennas 6 and 7. These are formed by symmetrically cutting a pair of edge slots and a center slot. The two edge slots and the center slot have the same locations as those of Antennas 3 and 5. The dimensions are $l_{es} = 60$ mm and $w_{es} = 2$ mm for the edge slots, and $l_{cs}, w_{cs} = 2$ mm for the center slot. The height H is 8 mm. The effects of changing the length of the center slot on the impedance characteristics of Antennas 6 and 7 are shown in Figure 3.77.

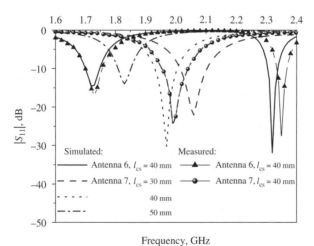

Figure 3.77 Measured and simulated return losses against frequency for Antennas 6 and 7 with various center slot lengths (Reproduced by permission of © IEEE. Z. N. Chen, 'Suspended plate antennas with shorting strips and slots,' *IEEE Transactions on Antennas and Propagation*, vol. 52, no. 10, pp. 2525–2531, October 2004.)

Similar to Antennas 4 and 5, Antennas 6 and 7 also operate in a dual-band mode. The lower resonant frequency of Antenna 7 decreases from 2.06 GHz to 1.80 GHz as the center slot is extended from 30 mm to 60 mm. The impedance matching worsens for the length $l_{cs} > 40$ mm. The simulated and measured bandwidths ($|S_{11}| < -10$ dB) for the lower bands are 3.5 %/2.9 % for Antenna 6 and 4.5 %/ 4.1 % for Antenna 7.

Figure 3.78 compares the impedance performances for all the SPAs discussed above. It is seen from the comparison that the shorting strips reduce the impedance bandwidth for $|S_{11}| < -10$ dB from 7 % to 5 %, and increase the resonant frequency by 5 %. However, the edge or center slots can partially offset the increase in the resonant frequency.

Radiation Characteristics

Since the radiation performance of broadband SPAs is usually deteriorated, one should carefully examine the radiation patterns at three typical operating frequencies, namely lower and upper edge frequencies (f_l, f_u), and the center frequency (f_r) of the bandwidth in the principal H- and E-planes.

Antennas 0 and 1. Figure 3.79 displays the radiation patterns at $f_r = 1.92$ GHz for Antenna 0 and 2.11 GHz for Antenna 1 in the E- and H-planes. The measured results show that within the bandwidth, the radiation patterns hardly change. In the E-plane, the co-polarized radiation patterns for Antenna 1 with the shorting strip are more symmetrical than those for Antenna 0. In the H-plane, the shorting strip does not affect the co-polarized radiation patterns. The measured results for Antennas 0, 1, 3, 5 and 7 are compared in Table 3.7.

For instance, in the H-plane, the co-to-cross-pol ratios for Antenna 1 (with a shorting strip) range from 2 dB to 4 dB above those for Antenna 0. However, in the E-plane,

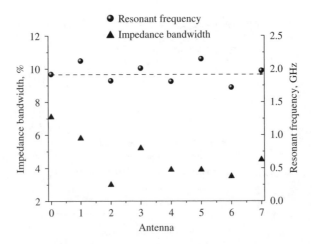

Figure 3.78 Comparison of impedance bandwidths and resonant frequencies between the SPAs. (Reproduced by permission of © IEEE. Z. N. Chen, 'Suspended plate antennas with shorting strips and slots,' *IEEE Transactions on Antennas and Propagation*, vol. 52, no. 10, pp. 2525–2531, October 2004.)

(a) H-plane

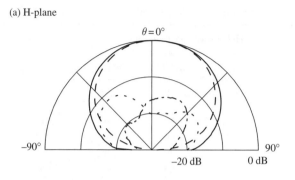

(b) E-plane

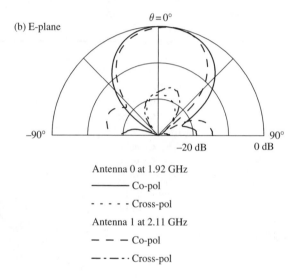

Antenna 0 at 1.92 GHz

———— Co-pol

- - - - - Cross-pol

Antenna 1 at 2.11 GHz

— — — Co-pol

—·—··Cross-pol

Figure 3.79 Measured radiation patterns for Antennas 0 and 1 at the resonant frequencies. (Reproduced by permission of © IEEE. Z. N. Chen, 'Suspended plate antennas with shorting strips and slots,' *IEEE Transactions on Antennas and Propagation*, vol. 52, no. 10, pp. 2525–2531, October 2004.)

the co-to-cross-pol levels of Antenna 1 decrease from 20.3 dB to 15.5 dB at 2.11 GHz. It should be noted that the resonant frequencies for Antenna 1 are 10 % higher than those for Antenna 0.

Antennas 2 and 3. Figure 3.80 shows the measured radiation patterns in the E- and H-planes at 1.97, 2.01 and 2.07 GHz for Antenna 3. As the slot is centrally cut on the plate, all the co-polarized radiation patterns are symmetrical. In the E-plane, the levels of two side lobes vary from −10.5 dB to −13.3 dB at around ±80°. Furthermore, Figure 3.81 shows the distributions of the imaginary part of the electric currents on the plates of Antennas 2 and 3 at the upper edge frequency f_u. For Antenna 2, the x-directed currents close to radiating edges are asymmetrical with respect to the y-axis and nearly in phase. In contrast, since the

Table 3.7 Comparison of measured results for Antennas 0, 1, 3, 5 and 7.

Antenna	Plane	Half-power beamwidth(deg)			Co-to-cross-pol ratio(dB)		
		At f_l	At f_c	At f_u	At f_l	At f_c	At f_u
0	E	60	58	52	19.8	18.8	21.8
0	H	76	74	75	12.5	12.3	11.5
1	E	52	50	46	20.3	16.4	15.5
1	H	67	66	65	16.5	14.3	14.3
3	E	50	50	50	28.0	24.4	17.7
3	H	66	66	66	13.6	16.9	16.5
5	E	47	45	43	17.2	15.3	16.3
5	H	68	67	67	16.9	16.0	15.4
7	E	46	46	45	16.8	16.7	17.2
7	H	65	65	65	24.0	28.6	20.4

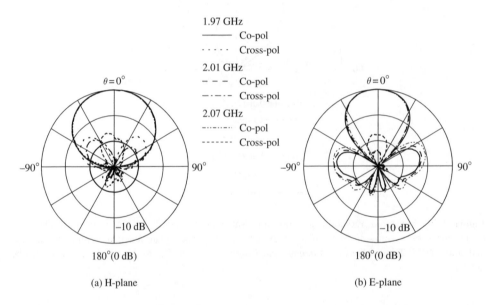

(a) H-plane   (b) E-plane

Figure 3.80 Measured radiation patterns for Antenna 3 at frequencies f_l, f_r, and f_u. (Reproduced by permission of © IEEE. Z. N. Chen, 'Suspended plate antennas with shorting strips and slots,' *IEEE Transactions on Antennas and Propagation*, vol. 52, no. 10, pp. 2525–2531, October 2004.)

current distribution of Antenna 3 at the upper edge frequency f_u is similar to that at the resonant frequency f_r, the radiation property of Antenna 3 remains quite satisfactory.

Antennas 4 and 5. Figure 3.82 depicts the measured radiation patterns for Antenna 5 in the E- and H-planes at 2.10, 2.15 and 2.20 GHz. Owing to the symmetrical location of the edge slots, the co-polarized radiation patterns are kept symmetrical in both E- and H-planes.

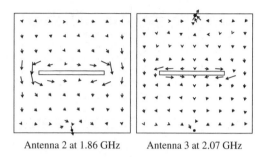

Antenna 2 at 1.86 GHz   Antenna 3 at 2.07 GHz

Figure 3.81 Simulated distributions of the imaginary parts of electric currents on Antennas 2 and 3 operating at the upper edge frequency. (Reproduced by permission of © IEEE. Z. N. Chen, 'Suspended plate antennas with shorting strips and slots,' *IEEE Transactions on Antennas and Propagation*, vol. 52, no. 10, pp. 2525–2531, October 2004.)

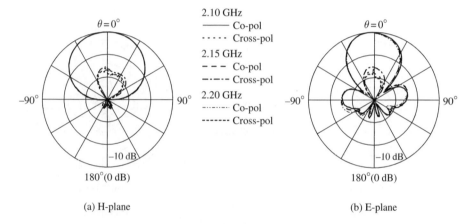

(a) H-plane   (b) E-plane

Figure 3.82 Measured radiation patterns for Antenna 5 at frequencies f_l, f_r and f_u. (Reproduced by permission of © IEEE. Z. N. Chen, 'Suspended plate antennas with shorting strips and slots,' *IEEE Transactions on Antennas and Propagation*, vol. 52, no. 10, pp. 2525–2531, October 2004.)

In the E-plane, the levels of the side lobes slightly change from -14.4 dB to -14.1 dB at around 74° and $-90°$. Table 3.7 details the effects of the edge slots and shorting strip on the half-power beamwidths and the co-to-cross-pol ratios. The results show that the edge slots meliorate the co-polarized radiation performance and suppress the cross-polarized radiation in the H-plane but increase cross-polarized radiation levels in the E-plane.

To understand the radiation characteristics of these SPAs, Figure 3.83 compares the distributions of the imaginary parts of the electric currents on Antennas 4 and 5 at the upper frequencies f_u. From the comparison, it is readily observed that the cross-directed currents on Antennas 4 and 5 mainly appear around the edge slots. The x-directed currents are symmetrical with respect to the y-axis but out of phase so that the x-polarized radiation can be cancelled out. In particular, the current distributions of Antenna 5 hardly change when the operating frequencies vary within the bandwidth. Thus, the radiation performance is quite stable across the bandwidth.

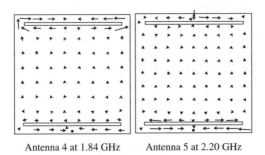

Antenna 4 at 1.84 GHz   Antenna 5 at 2.20 GHz

Figure 3.83 Simulated distributions of the imaginary parts of electric currents on Antennas 4 and 5 operating at the upper edge frequency. (Reproduced by permission of © IEEE. Z. N. Chen, 'Suspended plate antennas with shorting strips and slots,' *IEEE Transactions on Antennas and Propagation*, vol. 52, no. 10, pp. 2525–2531, October 2004.)

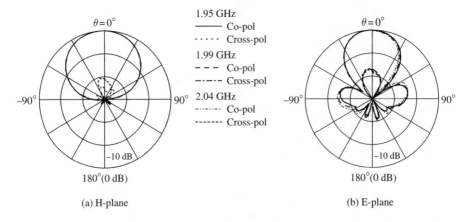

(a) H-plane   (b) E-plane

Figure 3.84 Measured radiation patterns for Antenna 7 at frequencies f_l, f_r, and f_u. (Reproduced by permission of © IEEE. Z. N. Chen, 'Suspended plate antennas with shorting strips and slots,' *IEEE Transactions on Antennas and Propagation*, vol. 52, no. 10, pp. 2525–2531, October 2004.)

Antennas 6 and 7. The measured radiation patterns for Antenna 7 in the E- and H-planes at 1.95, 1.99 and 2.04 GHz are illustrated in Figure 3.84. The symmetrical structure makes the co-polarized radiation patterns in both E- and H-planes symmetrical. In the E-planes, the levels of the side lobes of the co-polarized radiation patterns vary from −15.2 dB to −14.4 dB around ±82°. Table 3.7 shows that the half-power beamwidths are quite stable in the E- and H-planes over the impedance bandwidth. The co-to-cross-pol ratios increase from 20.4 dB to 28.6 dB in the H-planes and vary about 17 dB in the E-planes. Evidently, the combination of the edge and center slots is able to enhance the radiation performance of these SPAs.

Finally, Figure 3.85(a) compares the stable half-power beamwidths in E- and H-planes and the unchanged average gain of the SPAs with shorting strips. Also, the co-to-cross-pol ratios for the SPAs are compared in Figure 3.85(b). The comparison shows that the introduction

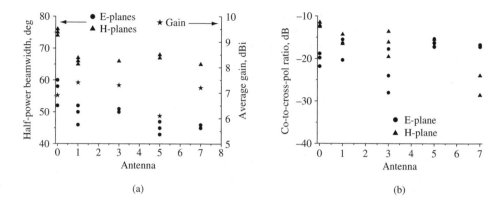

Figure 3.85 Comparison of (a) half-power beamwidths and average gain, and (b) co-to-cross-pol ratios. (Reproduced by permission of © IEEE. Z. N. Chen, 'Suspended plate antennas with shorting strips and slots,' *IEEE Transactions on Antennas and Propagation*, vol. 52, no. 10, pp. 2525–2531, October 2004.)

of the shorting strip and slots suppresses cross-polarized radiation in the H-plane at the expense of a slight increase in the cross-polarized radiation in the E-plane, although it is still kept low.

3.4 ARRAYS WITH SUSPENDED PLATE ELEMENTS

Owing to their broadband characteristics, suspended plate elements can be used to form broadband arrays. Inter-element mutual coupling should be taken into account in array design. This section discusses the mutual coupling between two suspended plate elements above a double-tiered ground plane. This technique is used to reduce the size of the array. The design idea is validated by designs with reduced lateral size. The compact design makes it suitable for adaptive arrays in cellular base stations.

The mutual coupling between the array elements significantly affects the performance of arrays in wireless communications applications. The mutual coupling in microstrip antenna arrays mainly attributes to space waves, higher-order waves, surface waves and leaky waves,[47] which significantly degrades the signal-to-interference-noise ratio (SINR) and direction-of-arrival (DOA) estimation in the case of an adaptive array.[48,49] Several theoretical and experimental investigations of mutual coupling have been reported.[50-54]

Typically, the mutual coupling may degrade radiation patterns in terms of increasing the side-lobe levels, shifting of nulls, and the appearance of grating lobes. Increasing the spacing between array elements can usually reduce or weaken the mutual coupling. However, the larger inter-element spacing results in a larger array size, with a higher installation cost because wireless system operators have to pay more for the space to install the array. Therefore, there is a need for a laterally compact antenna array configured appropriately to reduce the mutual coupling between array elements.

Reducing surface waves will aid in the suppression of mutual coupling. The solutions for reducing surface waves include increasing the separation between array elements with large size,[55] and removing the dense dielectric substrate from a microstrip patch antenna, which forms an SPA without any surface waves.[7] Recently, electromagnetic band-gap (EBG) structures have been proposed to reduce the mutual coupling between microstrip patch antennas.[56-58]. However, EBG structures worsen the mutual coupling in H-planes, although significant suppression of the mutual coupling in E-planes has been achieved.[56,57]

Some arrays have been constructed on nonplanar ground planes to conform with the installation environment. For example, an array may have curved surfaces for integration into aircraft and missiles.[59,60] Another design is pattern synthesis by use of a nonplanar array.[61]

This section of the chapter introduces methods of reducing an array's lateral size. Antenna arrays with tiered ground planes are constructed. Broadband two-element arrays with suspended plate radiators are first exemplified to show the mutual coupling between two radiators above the tiered ground planes. The impedance and radiation performance of the linear four-element arrays on the planar and tiered ground planes are compared numerically and experimentally.

3.4.1 MUTUAL COUPLING BETWEEN TWO SUSPENDED PLATE ELEMENTS

The array is installed on a tiered ground plane, which affects the inter-element mutual coupling, impedance and radiation performance of the array, as well as array size.

Figure 3.86 shows the geometry and coordinate system of the two-element array. The two square SPA elements ($L \times L = 70\,\text{mm} \times 70\,\text{mm}$) are located parallel to each other to

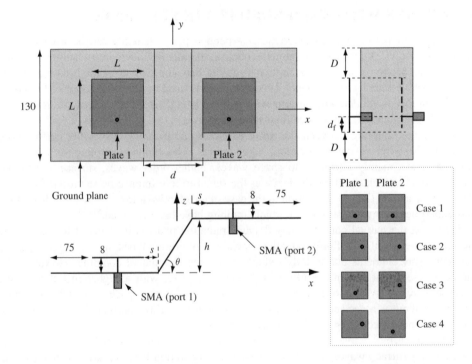

Figure 3.86 Geometry of a two-element array above a double-tiered ground plane (dimensions in millimeters).

form four configurations as shown in the diagram. The elements are positioned centrally above a ground plane at a height of 8 mm. The ground plane is 130 mm wide. The distance between the plates and the edges of the ground plane along the y-axis is D [$=(130 - L)/2 = 30$ mm]. The elements are fed vertically by two identical coaxial probes of radius of 0.625 mm through SMAs under the ground plane at the distance d_f. By changing d_f and the height of the plates, good impedance matching can be achieved with a broad bandwidth.

In this design, the operating frequency is around 1.83 GHz ($\lambda_o = 164$ mm). Between the two elements, a portion of the ground plane is inclined at an angle θ. The height difference of the ground planes is h. The horizontal distance between adjacent edges of two elements is d. Four cases are taken into account: Cases 1, 2 and 3/4 stand for the H-, E- and O-(orthogonal) plane configurations, respectively.

First, a prototype for Case 1 ($\theta = 90°$, $s = 16$ mm, $h = 84$ mm $\approx 0.5\lambda_o$, and $d = 2s \approx 0.2\lambda_o$) was measured. In the tests, the phase shift between two elements is 180° to compensate for the phase shift at the boresight caused by the height h. The comparison between simulated and measured S parameters was carried out by using a commercial EM simulator IE3D (Zeland) as shown in Figure 3.87. Within the 10-dB bandwidth of 5.5 %, the coupling between the elements is less than -27.2 dB. The maximum coupling also appears within the bandwidth. Compared to conventional microstrip antennas with the same horizontal distance, the mutual coupling between the SPAs is much lower.

Next, Figures 3.88–3.91 compare the mutual coupling with planar ground plane ($\theta = 0°$) and tiered ground plane ($\theta = 90°$) for the four cases, where the distance $d = 2s$ is varied and other parameters are the same as those used in Figure 3.88. As compared with the scenario with the planar ground plane, the elements installed above the tiered ground planes have much weaker H-plane mutual coupling, whereas the E-plane mutual coupling stays almost the same. However, it should be noted that the E-plane mutual coupling with the tiered ground plane is 10 dB lower than with the planar ground plane when the horizontal distance d reaches $0.1\lambda_o$. For the cases with tiered ground planes, the elements are out of sight of

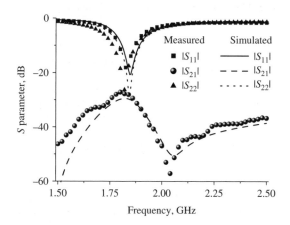

Figure 3.87 Comparison of the measured and simulated S parameters of the two-element array on a double-tiered ground plane.

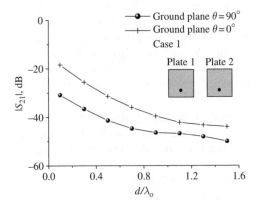

Figure 3.88 Comparison of the mutual coupling for Case 1.

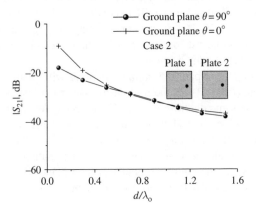

Figure 3.89 Comparison of the mutual coupling for Case 2.

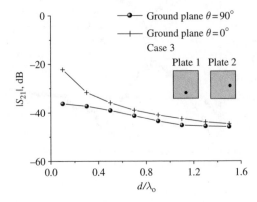

Figure 3.90 Comparison of the mutual coupling for Case 3.

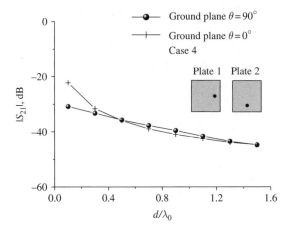

Figure 3.91 Comparison of the mutual coupling for Case 4.

each other and the distance between the elements is much larger than the horizontal distance d due to the vertical height $h \approx 0.5\lambda_0$. Therefore, for the tiered cases, the coupling due to space and higher-order waves becomes weaker than, or almost the same as, that for the cases with planar ground planes.

Among all the cases, the mutual coupling for Case 2 (E-plane) is the strongest. The mutual coupling for Case 1 (H-plane) greatly changes when the ground plane size is varied. The mutual coupling for both orthogonal cases (Cases 3 and 4) is lower than cases 1 and 2.

The effect of the height h on the mutual coupling for Cases 1 and 2 is shown in Figure 3.92. The mutual coupling for Case 1 (H-plane) is at least 10 dB lower than that for Case 2 (E-plane). With increasing height h, the mutual coupling for Case 1 decreases more rapidly than that for Case 2.

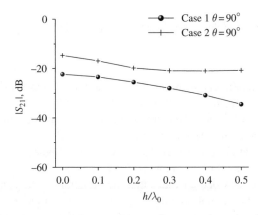

Figure 3.92 Effects of the height h on the mutual coupling for Cases 1 and 2.

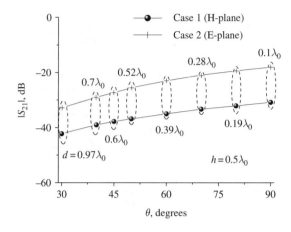

Figure 3.93 Effects of the inclined angle θ and distance d on the mutual coupling for Cases 1 and 2.

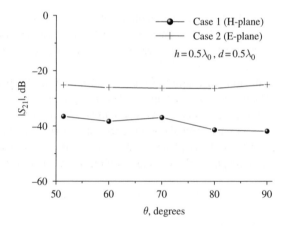

Figure 3.94 Effects of the inclined angle θ of the mutual coupling for Cases 1 and 2.

Figure 3.93 shows the effects of the inclined angle θ on the mutual coupling for Cases 1 and 2 when the height is fixed at $h = 84$ mm ($\approx 0.5\lambda_o$), and the distance s is 16 mm. As the angle θ increases from 30° to 90°, the coupling increases by about 10 dB for both cases with decrease in the distance d from $0.97\lambda_o$ to $0.1\lambda_o$.

Finally, the inclined angle θ is varied for Cases 1 and 2 when the distance between the edges of the SPAs is fixed with $h = 84$ mm ($\approx 0.5\lambda_o$) and $d = 84$ mm ($\approx 0.5\lambda_o$). The distance s is still 16 mm. Figure 3.94 demonstrates that the variation of the coupling is insignificant for Case 2. However, the mutual coupling for Case 1 with an angle $\theta = 90°$ is 3 dB lower than that of Case 1 with an angle $\theta < 70°$. Therefore, the tiered ground plane to some degree mitigates the mutual coupling between the suspended plate radiating elements in an array.

3.4.2 *REDUCED-SIZE ARRAY ABOVE DOUBLE-TIERED GROUND PLANE*

The foregoing discussion of mutual coupling of elements above a tiered ground plane has shown that the nonplanar ground plane does not degrade the performance in terms of the inter-element mutual coupling, although the horizontal distance d is quite small, for example $0.1\lambda_o$. This suggests that it is possible to reduce the lateral size of an array by using a tiered ground plane.

Figure 3.95 shows a linear array (Array 1) with four identical SPA elements above a tiered ground plane, and the coordinate system used in tests. The SPA elements are centrally deployed on a tiered ground plane and fed by 50-Ω coaxial probes through the ports R-1 to R-4 via SMAs. The array is designed to operate at 1.95 GHz (the wavelength λ_o is about 154 mm). The distance between two center elements (Elements 2 and 3) is around $\lambda_o/2$, whereas the horizontal separations between Elements 1 and 2 as well as Elements 3 and 4 are $0.13\lambda_o$ (20 mm) or 0.26 times the distance between the two center elements. This suggests a reduction in the lateral size of the array. The sections of the ground plane supporting Elements 1 and 4 are $\lambda_o/2$ higher than the center section of the ground plane. To compensate for the phase shift casued by the height difference

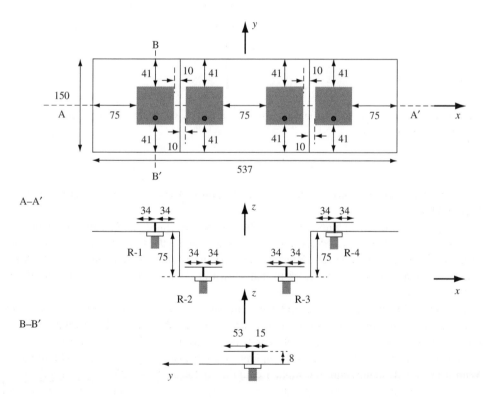

Figure 3.95 Geometry of the double-tiered four-element array with reduced lateral size (dimensions in millimeters).

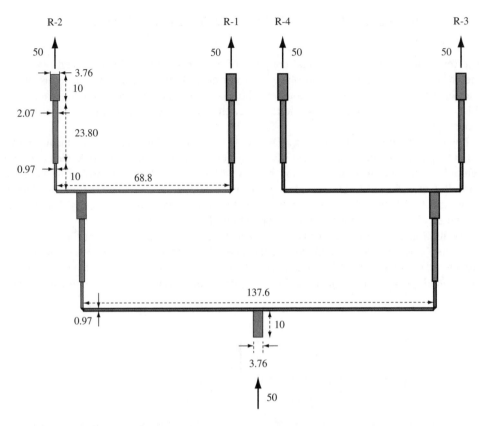

Figure 3.96 Structure of the feeding network for the double-tiered array with a 180° phase shift (dimensions in millimeters).

of the ground plane sections, a feeding network (power divider) was designed as shown in Figure 3.96.

A linear array (Array 2) with the same SPA elements operating at the same frequency but on a planar ground plane was designed as shown in Figure 3.97, to evaluate the performance of the tiered array (Array 1). The inter-element spacing is uniform and equal to $\lambda_o/2$. Array 2 is fed by a similar feeding network to that shown in Figure 3.96, but all four ports P-1 to P-4 are in phase.

The return losses for each element were measured with the other ports terminated by purely resistive matching loads of 50 Ω. The measurements showed good uniformity between the elements in each array. Figure 3.98 compares the results for P-1, R-1 and R-2.

The measured inter-element mutual coupling for the planar Array 2 is shown in Figure 3.99. Within the 10-dB bandwidth, the mutual coupling is lower than −25 dB. The results have very good agreement with those for the tiered Array 1 shown in Figure 3.100. This reveals that, despite the tiered ground plane, the array with reduced lateral size maintains almost the same mutual coupling levels as the planar array with a larger lateral size.

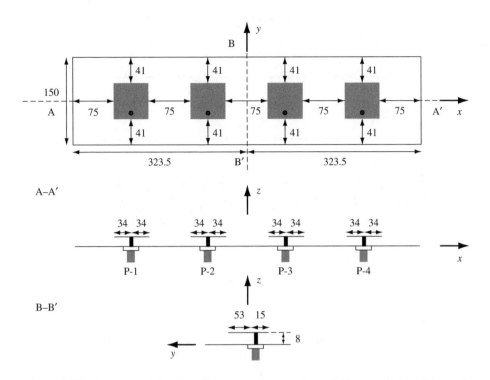

Figure 3.97 Geometry of the planar four-element array, Array 2 (dimensions in millimeters).

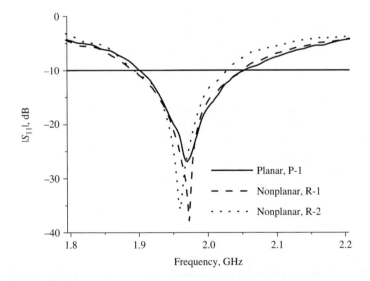

Figure 3.98 Comparison of the measured return losses between the planar and the double-tiered four-element arrays, Arrays 1 and 2.

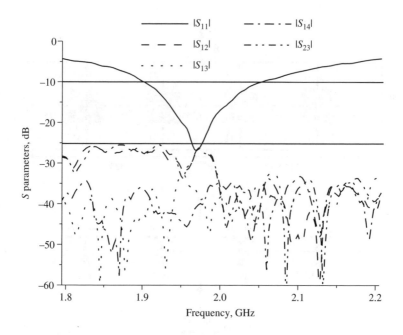

Figure 3.99   The inter-element mutual coupling for the planar Array 2.

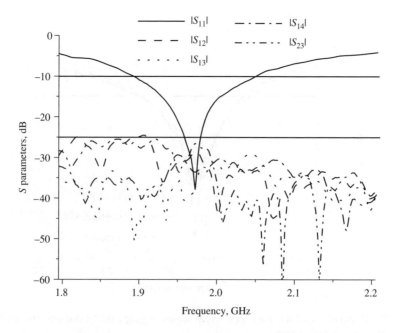

Figure 3.100   The inter-element mutual coupling for the planar Array 1.

The radiation patterns in both E- and H-planes across the bandwidths for Arrays 1 and 2 were compared. The radiation patterns at each frequency and plane have been normalized. The Arrays 1 and 2 were simulated and measured at lower and upper edge frequencies as well as at center frequencies. Figures 3.101 and 3.102 show the results for the planar Array 2.

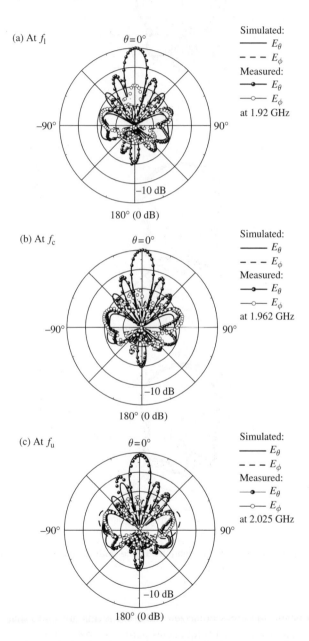

Figure 3.101 Comparison of simulated and measured radiation patterns for planar Array 2 at frequencies f_l, f_c, and f_u in the H-plane.

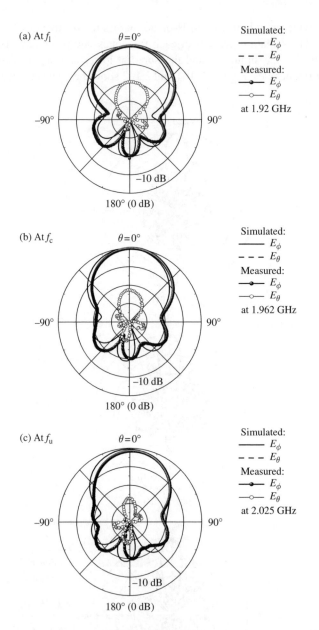

Figure 3.102 Comparison of simulated and measured radiation patterns for planar Array 2 at frequencies f_l, f_c, and f_u in the E-plane.

Within the main beam, the cross-polarized radiation levels are more than 20 dB below the co-polarized radiation levels at all frequencies across the bandwidth.

The radiation patterns for the double-tiered Array 1 are shown in Figures 3.103 and 3.104, which are very similar to Figures 3.101 and 3.102. The cross-polarized radiation levels are

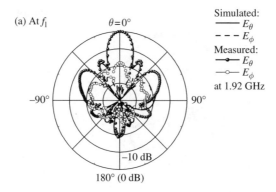

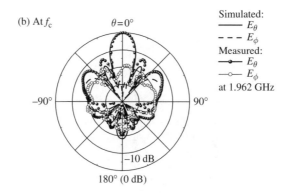

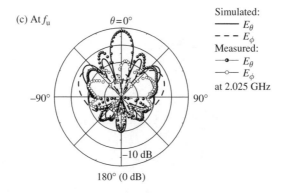

Figure 3.103 Comparison of simulated and measured radiation patterns for planar Array 1 at frequencies f_1, f_c and f_u in the H-plane.

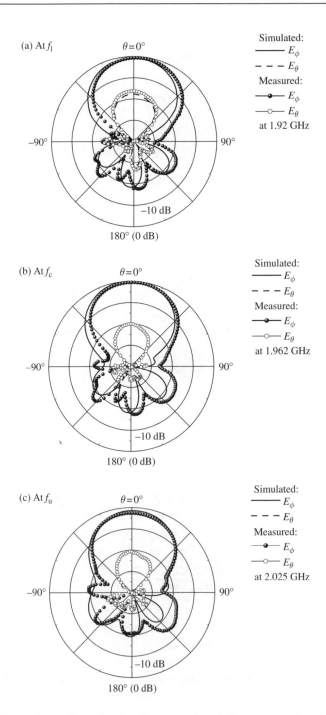

Figure 3.104 Comparison of simulated and measured radiation patterns for planar Array 1 at frequencies f_l, f_c and f_u in the E-plane.

maintained lower than -20 dB over the bandwidth. Compared to gain of 16.3 dBi for the planar Array 2 at the boresight, the tiered Array 1 has a gain of 14.8 dBi owing to the small ground plane in $x-y$ plane and higher side-lobe levels. In the H-plane, the half-power beamwidth of the main lobes of Array 1 (about $17°$) is $3°$ wider than that of Array 2 (about $14°$) because of the reduction in the size along the H-plane. In the H-plane, Array 2 has almost the same first side-lobe levels but higher second side-lobe levels than those of Array 1.

Next, a four-element linear tiered array (Array 3) was designed as shown in Figure 3.105. In this array, the two higher sections of the ground plane are connected to the lower section through an inclined ground plane. Array 3 is excited at the ports I-1 to I-4 by the feeding network shown in Figure 3.96. The measured return losses are almost the same as those for Arrays 1 and 2, and the mutual coupling between the elements at higher and lower sections of the ground plane is reduced owing to the larger separation compared with the tiered Array 1.

For comparison, Figures 3.106 and 3.107 show the radiation patterns in E- and H-planes. The radiation performance is quite similar to planar Array 2. The cross-polarized radiation levels at the boresight are also around 20 dB below co-polarized radiation levels.

Finally, a double-tiered array (Array 4) with the same configuration as the array shown in Figure 3.95 was excited by an anti-symmetrical feeding scheme. The configuration of the array is depicted in Figure 3.108. The adjacent elements form an anti-symmetrical feeding arrangement. Therefore, Array 4 is fed by the similar feeding network as the aforementioned planar Array 2, namely an in-phase feeding scheme for all ports S-1 to S-4. The return losses and mutual coupling for Array 4 are nearly the same as for the tiered Array 1. By means of

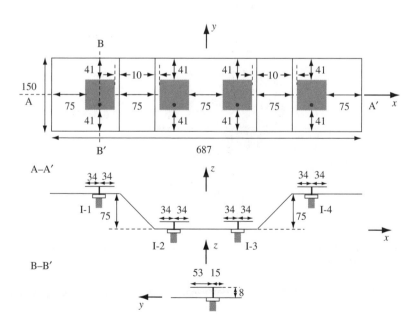

Figure 3.105 Geometry of a four-element linear double-tiered Array 3 (dimensions in millimeters).

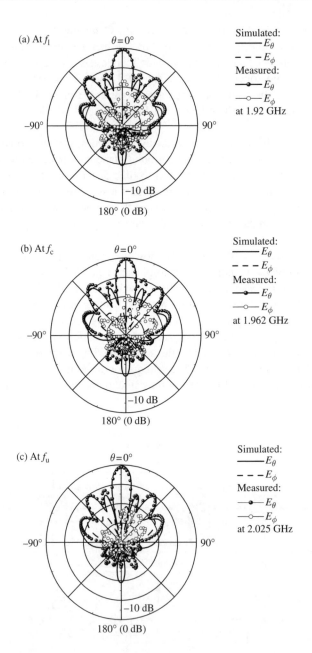

Figure 3.106 Comparison of simulated and measured radiation patterns for Array 3 at frequencies f_1, f_c and f_u in the H-plane.

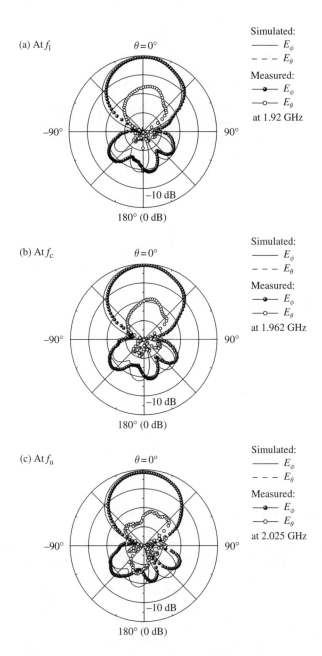

Figure 3.107 Comparison of simulated and measured radiation patterns for Array 3 at frequencies f_l, f_c and f_u in the E-plane.

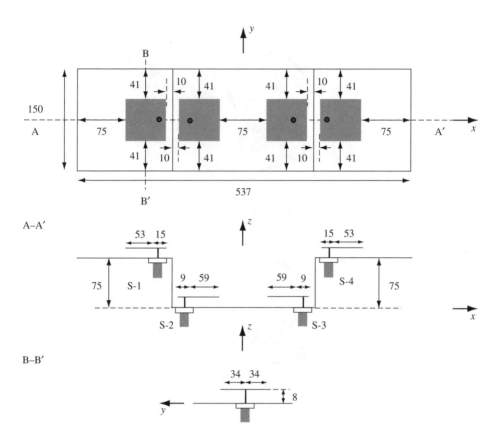

Figure 3.108 Geometry of four-element linear double-tiered Array 4 with an anti-symmetrical feeding scheme (dimensions in millimeters).

this feeding scheme, in theory, the cross-polarized radiation levels especially in H-planes are reduced to zero.

Figures 3.109 and 3.110 compare the simulated and measured radiation patterns in E- and H-planes for Array 4 over the bandwidth. It should be noted that, owing to the different feeding scheme, the radiation characteristics are different from the preceding three arrays. It is clear that the large difference between the simulated and measured cross-polarized radiation levels may be caused by the finite-size (small) ground plane and long RF feeding cable. Across the bandwidth the average measured gain of the array is 15.5 dBi.

In summary, arrays with suspended plate elements are able to achieve broad impedance bandwidths with low mutual coupling levels. To reduce the lateral size of arrays, a multi-tiered ground plane approach can be used. The comparison between arrays with a conventional planar ground plane and a double-tiered ground plane has shown that the latter, with a more than 60 % reduction in the inter-element spacing, has almost the same impedance and radiation performance as the conventional planar array. The array design can be used in base stations in wireless communication applications or phased arrays in radar systems.

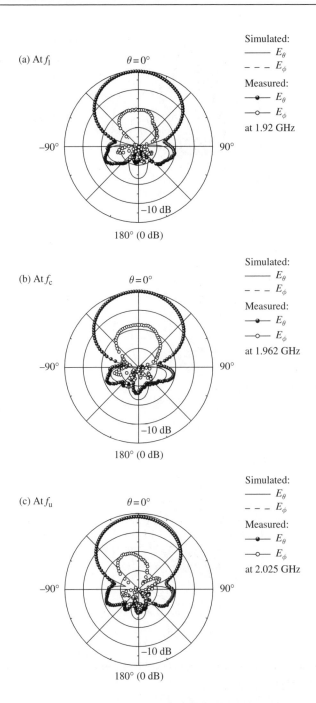

Figure 3.109 Comparison of simulated and measured radiation patterns for Array 4 at frequencies f_1, f_c and f_u in the H-plane.

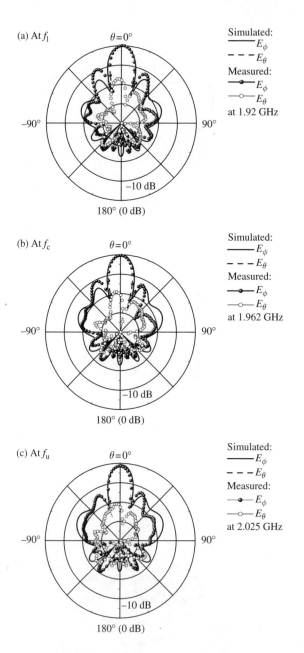

Figure 3.110 Comparison of simulated and measured radiation patterns for Array 4 at frequencies f_l, f_c and f_u in the E-plane.

REFERENCES

[1] N. Herscovici, 'A wide-band single-layer patch antenna,' *IEEE Transactions on Antennas and Propagation*, vol. 46, no. 3, pp. 471–473, 1998.

[2] Z. N. Chen and M. Y. W. Chia, 'Design of broadband probe-fed plate antenna with stub,' *IEE Proceedings: Microwave, Antennas and Propagation*, vol. 148, no. 4, pp. 221–226, 2001.

[3] J. M. Griffin and J. R. Forrest, 'Broadband circular disc microstrip antenna,' *Electronics Letters*, vol. 18, no. 5, pp. 266–269, 1982.

[4] K. F. Lee, K. Ho and J. Dahele, 'Circular-disk microstrip antenna with an air gap,' *IEEE Transactions on Antennas and Propagation*, vol. 32, no. 8, pp. 880–884, 1984.

[5] K. S. Fong, H. F. Guse and M. J. Withers, 'Wideband multiplayer coaxial-fed microstrip antenna element,' *Electronics Letters*, vol. 21, no. 11, pp. 497–499, 1985.

[6] G. Mayhew-Ridgers, J. W. Odendaal and J. Joubert, 'Single-layer capacitive feed for wideband probe-fed microstrip antenna elements,' *IEEE Transactions on Antennas and Propagation*, vol. 51, no. 6, pp. 1405–1407, 2003.

[7] T. Huynh and K. F. Lee, 'Single-layer single-patch wideband microstrip antenna,' *Electronics Letters*, vol. 31, no. 16, pp. 1310–1312, 1995.

[8] M. Clenet and L. Shafai, 'Multiple resonances and polarisation of U-slot patch antenna,' *Electronics Letters*, vol. 35, no. 2, pp. 101–103, 1999.

[9] K. F. Lee, K. M. Luk, K. F. Tong, Y. L. Yung and T. Huynh, 'Experimental study of a two-element array of U-slot patches,' *Electronics Letters*, vol. 32, no. 5, pp. 418–420, 1996.

[10] K. F. Lee, K. M. Luk, K. F. Tong, S. M. Shum, T. Huynh and R. Q. Lee, 'Experimental and simulation studies of the coaxially fed U-slot rectangular patch antenna,' *IEE Proceedings: Microwave, Antennas and Propagation*, vol. 144, no. 5, pp. 354–358, 1997.

[11] X. H. Yang and L. Shafai, 'Characteristics of aperture coupling microstrip antennas,' *IEEE Transactions on Antennas and Propagation*, vol. 43, no. 1, pp. 72–78, 1995.

[12] R. Bhalla and L. Shafai, 'Resonance behavior of single U-slot and double U-slot antenna,' *IEEE International Symposium on Antennas and Propagation*, vol. 2, pp. 700–703, 8–13 July 2001.

[13] S. Weigand, G. H. Huff, K. H. Pan and J. T. Bernhard, 'Analysis and design of broad-band single-layer rectangular U-slot microstrip patch antennas,' *IEEE Transactions on Antennas and Propagation*, vol. 51, no. 3, pp. 457–468, 2003.

[14] A. A. Deshmukh and G. Kumar, 'Half U-slot loaded rectangular microstrip antenna,' *IEEE International Symposium on Antennas and Propagation*, vol. 2, pp. 876–879, June 2003.

[15] J. Rosa, R. Nunes, A. Moleiro and C. Peixeiro, 'Dual-band microstrip patch antenna element with double U slots for gsm,' *IEEE International Symposium on Antennas and Propagation*, vol. 3, pp. 1596–1599, 2000.

[16] Z. N. Chen, M. Y. W. Chia and K. Hirasawa, 'Rectangular patch antenna with broad bandwidth,' *Millennium Conference on Antennas and Propagation*, vol. 1, Davos, Switzerland, 9–14 April 2000.

[17] Z. N. Chen, 'Experimental investigation on impedance characteristics of patch antenna with finite-size substrate,' *Microwave and Optical Technology Letters*, vol. 25, no. 2, pp. 107–111, 2000.

[18] Z. N. Chen, 'Experimental investigation on rectangular plate antenna with Ω-shaped slot,' *Radio Science*, vol. 36, no. 5, pp. 833–840, 2001.

[19] Z. N. Chen and M. Y. W. Chia, 'Broadband rectangular slotted plate antenna,' *IEEE International Symposium on Antennas and Propagation, Salt Lake City, Utah, USA*, vol. 1, pp. 640–643, July 2000.

[20] H. Nakano, M. Yamazaki and J. Yamauchi, 'Electromagnetically coupled curl antenna,' *Electronics Letters*, vol. 33, no. 12, pp. 1003–1004, 1997.

[21] K. M. Luk, C. L. Mak, Y. L. Chow and K. F. Lee, 'Broadband microstrip antenna,' *Electronics Letters*, vol. 34, no. 15, pp. 1442–1443, 1998.

[22] A. Kishk, K. F. Lee, W. C. Mok and K. M. Luk, 'A wide-band small size microstrip antenna proximately coupled to a hook shape probe,' *IEEE Transactions on Antennas and Propagation*, vol. 52, no. 1, pp. 59–65, 2004.

[23] Z. N. Chen and M. Y. W. Chia, 'Broadband suspended plate antenna with probe-fed strip,' *IEE Proceeedings: Microwave, Antennas and Propagation*, vol. 148, no. 1, pp. 37–40, 2001.

[24] Z. N. Chen and M. Y. W. Chia, 'Broadband probe-fed plate antenna,' *30th IEEE Europe Microwave Conference*, vol. 1, pp. 182–185, Paris, 2–6 October 2000.

[25] Z. N. Chen, K. Hirasawa and K. Wu, 'A broadband sleeve monopole integrated into a parallel-plate waveguide,' *IEEE Transactions on Microwave Theory and Techniques*, vol. 48, no. 7, pp. 1160–1163, 2000.

[26] Y. Kim, N. J. Farcich, S. Kim, S. Nam and P. M. Asbeck, 'Size reduction of microstrip antenna by elevating centre of patch,' *Electronics Letters*, vol. 38, no. 20, pp. 1163–1165, 2002.

[27] Y. Kim, G. Lee and S. Nam, 'Efficiency enhancement of microstrip antenna by elevating radiating edges of patch,' *Electronics Letters*, vol. 38, no. 19, pp. 1363–1364, 2002.

[28] Z. N. Chen, 'Broadband suspended plate antennas with concaved center portion,' *IEEE Transactions on Antennas and Propagation*, vol. 53, no. 4, pp. 1550–1551, April 2005.

[29] Z. N. Chen and M. Y. W. Chia, 'A feeding scheme for enhancing impedance bandwidth of a suspended plate antenna,' *Microwave and Optical Technology Letters*, vol. 38, no. 1, pp. 21–25, 2003.

[30] Z. N. Chen and M. Y. W. Chia, 'Experimental study on radiation performance of probe-fed suspended plate antenna,' *IEEE Transactions on Antennas and Propagation*, vol. 51, no. 8, pp. 1964–1971, 2003.

[31] T. Chiba, Y. Suzuki and N. Miyano, 'Suppression of higher modes and cross-polarized component for microstrip antennas,' *IEEE International Symposium on Antennas and Propagation*, vol. 1, pp. 285–288, June 1982.

[32] J. Huang, 'A technique for an array to generate circular polarization with linearly polarized elements,' *IEEE Transactions on Antennas and Propagation*, vol. 34, no. 9, pp. 1113–1124, 1986.

[33] P. S. Hall, 'Probe compensation in thick microstrip patches,' *Electronics Letters*, vol. 23, pp. 606–607, 1987.

[34] H. D. Foltz, J. S. McLean and G. Crook, 'Disk-loaded monopoles with parallel strip elements,' *IEEE Transactions on Antennas and Propagation*, vol. 46, no. 9, pp. 1894–1896, 1998.

[35] K. Hirasawa and M. Haneishi (Ed.), *Analysis, Design, and Measurement of Small and Low-profile Antennas*. Boston, MA: Artech House, 1991.

[36] J. J. Schuss and R. L. Bauer, 'Axial ratio of balanced and unbalanced fed circularly polarized patch radiator arrays,' *IEEE International Symposium on Antennas and Propagation*, vol. 1, pp. 286–289, June 1987.

[37] A. Petosa, A. Ittipiboon and N. Gagnon, 'Suppression of unwanted probe radiation in wideband probe-fed microstrip patches,' *Electronics Letters*, vol. 35, pp. 355–357, 1999.

[38] W. Hsu and K. Wong, 'A dual capacitively fed broadband patch antenna with reduced cross-polarization radiation,' *Microwave and Optical Technology Letters*, vol. 26, no. 3, pp. 169–171, 2000.

[39] K. Levis, A. Ittipiboon and A. Petosa, 'Probe radiation cancellation in wideband probe-fed microstrip arrays,' *Electronics Letters*, vol. 36, no. 7, pp. 606–607, 2000.

[40] Z. N. Chen and M. Y. W. Chia, 'Broadband suspended probe-fed plate antenna with low cross-polarization levels,' *IEEE Transactions on Antennas and Propagation*, vol. 51, no. 2, pp. 345–346, 2003.

[41] Z. N. Chen and M. Y. W. Chia, 'A novel center-fed suspended plate antenna,' *IEEE Transactions on Antennas and Propagation*, vol. 51, no. 6, pp. 1407–1410, 2003.

[42] Z. N. Chen and J. H. Ng, 'Probe-fed center-slotted suspended plate antennas with resistive and capacitive loadings,' *Microwave and Optical Technology Letters*, vol. 45, no. 4, pp. 355–360, May 20, 2005.

[43] Z. N. Chen and M. Y. W. Chia, 'Broadband suspended plate antennas fed by double L-shaped strips,' *IEEE Transactions on Antennas and Propagation*, vol. 52, no. 9, pp. 2496–2500, 2004.

[44] R. B. Waterhouse, 'Small printed antennas with low cross-polarised fields,' *Electronics Letters*, vol. 33, no. 15, pp. 1280–1281, 1997.

[45] P. P. S. Maci, G. B. Gentili and C. Salvador, 'Dual-band slot-loaded patch antenna,' *IEE Proceedings: Microwave, Antennas and Propagation*, vol. 142, no. 3, pp. 225–232, 1995.

[46] T. S. P. See and Z. N. Chen, 'Slot-loaded center-fed microstrip patch antenna,' *Microwave and Optical Technology Letters*, vol. 34, no. 3, pp. 227–232, 2002.

[47] N. G. Alexopoulos and I. E. Rana, 'Mutual impedance computation between printed dipoles,' *IEEE Transactions on Antennas and Propagation*, vol. 29, no. 1, pp. 106–111, 1981.

[48] I. Gupta and A. Ksienski, 'Effect of mutual coupling on the performance of adaptive arrays,' *IEEE Transactions on Antennas and Propagation*, vol. 31, no. 5, pp. 785–791, 1983.

[49] K. M. Pasala and E. M. Friel, 'Mutual coupling effects and their reduction in wideband direction of arrival estimation,' *IEEE Transactions on Aerospace Electronic Systems*, vol. 30, no. 4, pp. 1116–1122, 1994.

[50] D. M. Pozar, 'Input impedance and mutual coupling of rectangular microstrip antennas,' *IEEE Transactions on Antennas and Propagation*, vol. 30, no. 6, pp. 1191–1196, 1982.

[51] E. Penard and J. P. Daniel, 'Mutual coupling between microstrip antennas,' *Electronics Letters*, vol. 18, no. 14, pp. 605–607, 1982.

[52] R. P. Jedlicka, M. T. Poe and K. R. Carver, 'Measured mutual coupling between microstrip antennas,' *IEEE Transactions on Antennas and Propagation*, vol. 29, no. 1, pp. 147–149, 1981.

[53] G. Dubost, 'Influence of surface wave upon efficiency and mutual coupling between rectangular microstrip antennas,' *IEEE International Symposium on Antennas and Propagation*, vol. 2, pp. 660–663, 1990.

[54] R. Q. Lee, T. Talty and K. F. Lee, 'Measured mutual coupling between two-layer electromagnetically coupled patch antennas,' *IEEE International Symposium on Antennas and Propagation*, vol. 3, pp. 1364–1367, July 1991.

[55] M. A. Khayat, J. T. Williams and S. A. Long, D. R. Jackson, 'Mutual coupling between reduced surface-wave microstrip antennas,' *IEEE Transactions on Antennas and Propagation*, vol. 48, no. 10, pp. 1581–1593, 2000.

[56] F. Yang and Y. Rahmat-Samii, 'Microstrip antennas integrated with electromagnetic band-gap structure: a low mutual coupling design for array applications,' *IEEE Transactions on Antennas and Propagation*, vol. 51, no. 10, pp. 2936–2946, 2003.

[57] Z. Iluz, R. Shavit and R. Bauer, 'Microstrip antennas phased array with electromagnetic band-gap structure,' *IEEE Transactions on Antennas and Propagation*, vol. 51, no. 6, pp. 1446–1453, 2003.

[58] L. Zhang, J. A. Castaneda and N. G. Alexopoulos, 'Scan blindness free phased array design using pbg materials,' *IEEE Transactions on Antennas and Propagation*, vol. 52, no. 8, pp. 2000–2007, 2004.

[59] P. H. Pathak and N. Wang, 'Ray analysis of mutual coupling between antennas on a convex surfaces,' *IEEE Transactions on Antennas and Propagation*, vol. 29, no. 6, pp. 911–922, 1981.

[60] W. Y. Tam, A. K. Y. Lai and K. M. Luk, 'Mutual coupling between cylindrical rectangular microstrip antennas,' *IEEE Transactions on Antennas and Propagation*, vol. 43, no. 8, pp. 897–889, 1995.

[61] N. Herscovici, 'Nonplanar microstrip arrays,' *IEEE Transactions on Antennas and Propagation*, vol. 44, no. 3, pp. 389–392, 1996.

4

Planar Inverted-L/F Antennas

4.1 INTRODUCTION

Mobile and wireless communication systems are extremely popular, some device examples being cellular phones, laptop computers, PDAs, in-building access points of WLANs and GPS terminals. Wireless devices for personal or vehicular applications need to be portable, so they must be small in size and light in weight. The antenna for these devices has been the largest component except for the battery and display. As a result, there is a strong demand for reducing the size of antennas.

An antenna with a low profile is one type of miniaturized design that is widely applied in vehicle-mounted devices. To reduce the height of a conventional wire monopole, the top end of the wire monopole is folded such that the monopole consists of two radiating segments, one vertical and the other horizontal. However, with the reduction in the height of the monopole, its operating mode changes from series resonance, whereby the monopole length is about a quarter of the operating wavelength, to parallel resonance, whereby the overall monopole length is about half the operating wavelength. A quarter-wavelength thin-wire monopole usually has half the impedance of a half-wavelength dipole, namely $(73 + j42.5)/2\,\Omega$, and operates around its first resonance. To change the length and/or thickness of the monopole, the imaginary part will be cancelled out and the pure resistive impedance will be around $50\,\Omega$. It is easy to match the monopole to transmission systems such as waveguides, coaxial cables, striplines, microstrip lines and twin-lead cables because the characteristic impedance of industry-standard transmission systems is typically $50\,\Omega$, $75\,\Omega$ or $100\,\Omega$. Sometimes, an impedance matching transformer is used in the antenna to achieve the best matching between the antenna and feeding structures.

In order to achieve good impedance matching, a shorting pin can be introduced close to the vertical segment so that a shunt-driven inverted-L antenna transmission with an open end – an inverted-F antenna – is formed.[1,2] However, the impedance bandwidth of

Broadband Planar Antennas: Design and Applications Zhi Ning Chen and Michael Y. W. Chia
© 2006 John Wiley & Sons, Ltd

an inverted-F antenna is narrow owing to the high Q value. To broaden the bandwidth of the thin-wire inverted-F antenna for existing wireless devices, many techniques have been developed.

A simple method is to replace the thin-wire radiating segment of an inverted-F/L antenna with a wide planar radiator. The antenna is then called a *planar inverted-F/L antenna* (PIFA/PILA). Typically, a PIFA/PILA above an infinite ground plane has an impedance bandwidth of up to 2–3 %, compared with less than 1 % for a thin-wire inverted-F/L antenna in its basic form. Built-in PIFAs in detachable handsets were first reported in 1984 for GSM 800-MHz band mobile phones.[3] In-depth analyses of the performance and characteristics have been carried out.[4,5]

With the rapid miniaturization of wireless terminals and the increase in usage of broadband systems, broadband PIFAs/PILAs have been the major candidates for a variety of portable devices. However, the bandwidths of 2–3 % are not enough for many applications. For example, for the GSM 900-MHz band, a bandwidth of 8.7 % is necessary (see Table 1.1). Accordingly, techniques to broaden the impedance bandwidths of PIFAs/PILAs have been developed. The following are the commonly used techniques in practice:

- Replace the shorting wire by a shorting strip.
- Add parasitic radiators coupled to the driven element.[6–12]
- Modify the feeding structure.[13–15]
- Cut slot(s) from or notch the radiator.[16–20]
- Load the radiators with resistive, capacitive, inductive or dielectric material loading.[21–26]
- Use a combination of the techniques mentioned above.[27,28]

It should be mentioned that in many practical applications, systems often operate in multiple modes at different frequencies; for instance, mobile phones usually work in dual or more bands. In European countries, mobile phones cover at least the GSM900 and GSM1900 bands as mentioned in Table 1.1, where each band has a broad operating bandwidth. To cater for international roaming, sometimes mobile phones must cover some of the bands for mobile communications as listed in Table 1.1, so dual-band, triple-band and even multi-band designs are desirable. Furthermore, the mobile phone may operate in dual modes with additional capability to access, for instance, WLANs at the ISM band compliant with IEEE Standards 802.11a/b/g. Some portable devices, such as mobile phones, PDAs and laptop computers, may cover all the bands. Other devices used for vehicles, mobile phones or laptop computers need to use GPS.[29] Recently, more critical requirements for portable devices covering the UWB band and other bands for cellular phones and WLAN applications have been discussed. Therefore, multi-band designs with broad sub-bands are considered to be a type of broadband design.

In this chapter, the basic concept of inverted-F/L antennas is introduced. Next, some important characteristics of PIFAs/PILAs are described, and parametric studies of the antennas are carried out. After that, some techniques for the enhancement of impedance bandwidth are briefly reviewed, and two broadband PILAs, L-shaped and notched, are exemplified. Then, the design of a PILA with a vertical ground plane is investigated. Finally, two designs are highlighted as case studies: a PIFA for handphone applications, and the other for laptop computer applications.

4.2 THE INVERTED-L/F ANTENNA

The straight wire monopole is the antenna with the most basic form. Its dominant resonance appears at around one-quarter of the operating wavelength, where its input impedance is nearly purely resistive and has a resistance of $\sim 37\,\Omega$, which matches well to the 50-Ω characteristic impedance of a cable. Monopoles and their variants have been commonly used. However, the height of quarter-wavelength has restricted their applications where a low-profile design is necessary, such as missiles, aircraft, space shuttles and vehicles. The rapid development of miniature electronic components, integrated chips and even batteries has meant that device sizes have shrunk dramatically. For instance, the volume of a handphone body has been reduced from thousands of cubic centimeters, for voice transmission only, to less than $100\,\text{cm}^3$ with multimedia functions – see Figure 4.1. The antennas have dwindled from more than 200-mm helixes to zero (because some of them have been fully embedded into the casing). However, the majority of handphone antennas are still monopoles, but with improved versions – namely the inverted-L or -F antennas.

In this section, inverted-L/F antennas in their basic forms are analysed experimentally.[30] First, the effect of lowering the profile of the monopole on the input impedance is investigated. Next, the effect of introducing an additional element on the impedance matching is addressed. The simulations are carried out using the EM simulator IE3D (Zeland).

Figure 4.2 shows the geometry of a narrow-strip monopole with a horizontal bent portion. A 4-mm wide strip monopole antenna is vertically mounted on a perfectly electrically conducting (PEC) plate and fed by a 50-Ω SMA. The overall length $(l+h)$ of the monopole is 35 mm. The frequency range in the tests is from 1 GHz to 5 GHz.

Figure 4.3 shows the input impedance for varying ratios h/l. As the ratio increases, the peak of input impedance significantly decreases. For example, the peak resistance decreases from 1407 Ω for $h=5$ mm to 293 Ω for $h=35$ mm. The monopoles with lower profiles have higher Q values, which makes impedance matching difficult and results in a narrow impedance

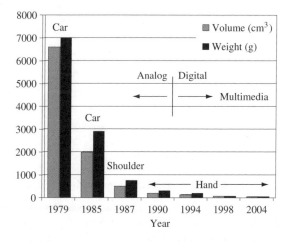

Figure 4.1 Change in the volume and weight of portable phones in the past 25 years. (The data were kindly provided by Professor Kazihiro Hirasawa, University of Tsukuba, Japan.)

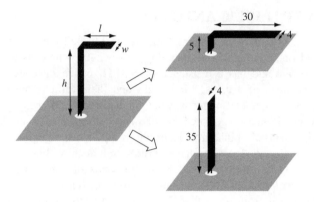

Figure 4.2 Geometry of a monopole with a horizontal bent portion (dimensions in millimeters).

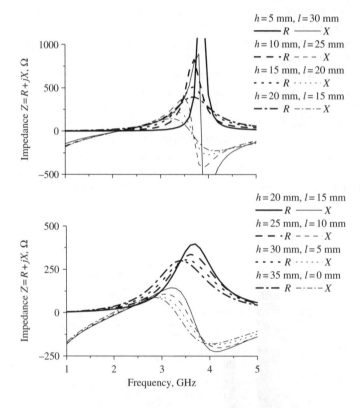

Figure 4.3 Input impedance for varying height h.

bandwidth. On the other hand, with increasing height h, the first resonant frequencies stay at around 2.1 GHz with increasing resistance from 3.0 Ω for $h = 5$ mm to 37.1Ω for $h = 35$ mm, whereas the second resonant frequencies decrease from 3.8 GHz for $h = 5$ mm to 3.3 GHz for $h = 35$ mm.

In addition, when h is lower than about $0.1\lambda_o$ the resistance R increases quickly and the radiation characteristics are significantly different from those of a straight monopole.[30] Therefore, $0.1\lambda_o$ is the critical height below which the antenna can be considered as an ILA rather than a top-loaded monopole antenna. This conclusion is the same as that discussed in Chapter 1, where planar monopoles were investigated in a similar way.

Furthermore, from Figure 4.4 it can been seen that a good impedance matching can be achieved when the ratio h/l is greater than 4/3, and the impedance matching becomes better with increasing h/l. On the other hand, the impedance matching becomes poorer as the profile of the monopole becomes lower. To maintain good impedance matching, a parasitic L-element is introduced at the feed point, which forms an IFA. As shown in Figure 4.5, the L-shaped element with the same width of 4 mm and height of 5 mm as the ILA is electrically attached to the ILA antenna. The vertical portion of the L-shaped element is grounded and its horizontal portion has a length d.

Figure 4.6 shows the input impedance of the IFA for varying distance d from 0 mm to 5 mm. A parallel resonance appears around 2 GHz when d is not zero, which is half the frequency at which the IFA resonants in parallel format. The simulation shows that the IFA achieves good impedance matching around 2 GHz when d is about 1.5 mm. This observation has been verified in microstrip antennas, where the introduction of a shorting wall at the center of the patch reduces the size of the original patch. Essentially, ILAs or IFAs and microstrip antennas are all open-ended transmission-line type antennas.

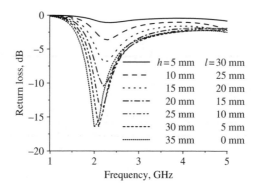

Figure 4.4 Impedance mismatch for varying height h.

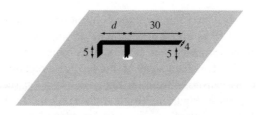

Figure 4.5 Geometry of the PIFA (dimensions in millimeters).

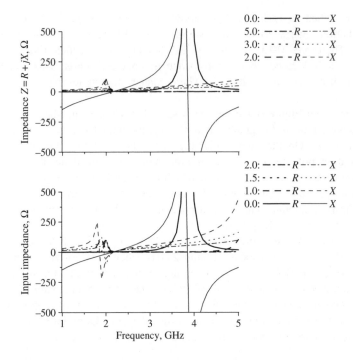

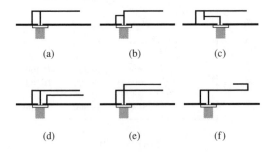

Figure 4.6 Input impedance for varying distance d (in millimeters).

Figure 4.7 Some variations of conventional inverted-F antennas.

In practical designs, an IFA may not be in the conventional form shown in Figure 4.7(a). Variations of conventional IFAs allow opportunities to improve the impedance matching and/or broad/multi-band performance. For an IFA in its basic form, the freedom for design mainly includes the parasitic L-shaped element, the feeding structure and the long radiating arm. Therefore, variations can arise from changing the location (height) of the parasitic L-shaped element, as shown in Figure 4.7(b), and changing the feeding stem as shown in Figure 4.7(c). Both of these methods allow good impedance matching. Additionally, the radiating arm can be modified as shown in Figures 4.7(d)–(f).[8,12]

In Figure 4.7(d), a parasitic L-shaped structure is electrically coupled to the radiating arm to form the second radiator. The smaller size of the second radiator may generate a second resonance at a higher frequency band. The parasitic radiator can be at any position around the main radiating arm that gives enough coupling to the main arm. Alternatively, the second radiator can be connected to the main arm or feeding stem directly, as depicted in Figure 4.7(e). Similarly, it can be placed at any position; for example, it can be parallel to or right under the main radiating arm. Figure 4.7(f) exhibits a variation in which the radiating arm is folded, so the length of the antenna is reduced. The folded part can be more complicated than the one shown in (f): it may be multi-folded and situated in any plane.

4.3 BROADBAND PLANAR INVERTED-F/L ANTENNA

Replacing the thin-wire radiator of an ILA or IFA with a planar radiator has been widely used to broaden the bandwidth and decrease the resonant frequency. A merit of this type of antenna is that it has more design freedom than a thin-wire one. As like the case for a microstrip antenna, planar radiators in PILAs or PIFAs can be reshaped by slitting slots or notches on them. A slit radiator has either reduced size or multiple well-matched bandwidths.

4.3.1 PLANAR INVERTED-F ANTENNA

Basically, a PIFA can be considered as a modification of a monopole or a microstrip patch antenna, as shown in Figure 4.8. Figure 4.8(a) shows the development of a thin-wire monopole into a low profile PIFA. More directly, a PIFA can originate from a planar monopole by bending the planar radiator for a low profile and introducing a shorting pin for good impedance matching, as shown in (b). Alternatively, the PIFA can be a variation

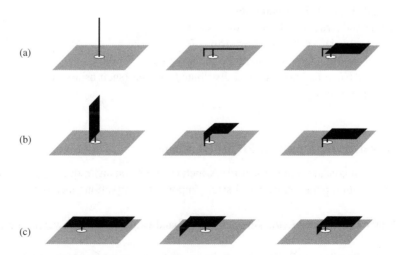

Figure 4.8 Development of PIFA from monopole, planar monopole and microstrip patch antennas.

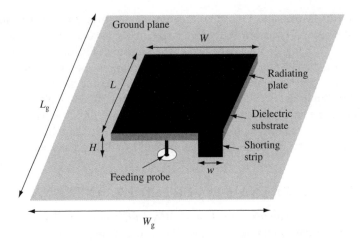

Figure 4.9   Geometry of a planar inverted-F antenna.

of a shorted patch antenna, where the radiating patch of the antenna is halved at its midline by a short-circuiting wall.[31,32,5] If the width of the shorting wall is further reduced to a narrow shorting strip, the PIFA is formed, as in (c). Therefore, the PIFA can be regarded as the variation of either a thin-wire transmission-line ILA or a short-circuit microstrip patch antenna.

Figure 4.9 shows a typical PIFA above a finite ground plane. The performance of this PIFA can be expressed in terms of the following parameters:

- the geometric shape of the radiating plate – L, W
- the height of the radiating plate – H
- the size and shape of the ground plane – L_g, W_g
- the location and structure of the feeding stem
- the location and size of the shorting strip – H, w
- the material used to support or load the radiating plate, if any
- lump loading (such as an LCR) or a distributed loading (such as a slot or a notch at the radiating plate, if any).

Geometric Parameter Effects on PIFA Performance

The main design considerations for a PIFA include the resonant frequency, the impedance bandwidth, radiation patterns, gain and size.[5] Important observations include:

- The greater the height H, the broader is the bandwidth, and the lower is the resonant frequency.
- The greater the ratio, w/W is (≤ 1), the higher is the resonant frequency, and the broader is the bandwidth.

- The greater the ratio of W/L, the lower is the resonant frequency, and the broader is the bandwidth.
- The locations of feed point and shorting strip, as well as the width w of the shorting strip, control the radiation polarization characteristics.

Knowing the effects of geometric parameters on the performance is essential for antenna designers when optimizing a PIFA for a specific application. Most importantly, the antenna designer should be aware of the effects on bandwidth and resonant frequency. In practice, an EM simulator can hardly provide exact design parameters, owing to complicated installation environment and fabrication tolerances. Fine adjustments of the parameters are usually necessary in practical antenna design.

Ground Plane Effects on PIFA Performance

One important factor that significantly affects performance is the ground plane.[33] This includes its size, shape and properties. At times, the ground plane is not electrically large and flat; examples are PIFAs used in portable devices such as PDAs, handsets and laptop computers. It should be noted that the ground plane affects not only PIFA performance but also the performance of other kinds of antenna, such as monopoles, dipoles, PILAs and microstrip antennas.[34–40] Essentially, the ground plane is part of the antenna.

Size. The size of the ground plane to a great extent affects the resonant frequency and radiation patterns when a PIFA is used in portable devices such as handsets and PDAs, where the ground plane is smaller than or comparable to the operating wavelength. It has been suggested that the ground plane size has a significant effect on the performance of the PIFA installed at the middle of a square ground plane when the size of ground plane is less than 0.2 times the operating wavelength. The reduced-size ground plane results in a higher resonant frequency, a narrower impedance bandwidth, as well as lower directivity.[33,41] However, the performance of the PIFA is strongly dependent on the orientation and position of the PIFA at the finite-size ground plane. For instance, a PIFA placed around the corner of a finite-sized ground plane has a broader impedance bandwidth and higher gain.[33,42]

In addition, a finite ground plane leads to the difficulty in making accurate measurements. Usually, an RF cable grounded to the measurement equipment must be connected to the ground plane of the antenna under test. When the ground plane is electrically large, the effect of the RF cable is ignored because the shielding effect of the ground plane suppresses the currents on the outer surface of the RF cable. However, the cable effect is significant on the measured results – especially the resonant frequency, radiation patterns and gain – when the size of the ground plane is comparable to an operating wavelength. Some techniques have been developed to alleviate the impact of the cable on the measurements.[43–46]

Shape. The shape of the ground plane where a PIFA is installed should be taken into account. One scenario of interest is the PIFA used in a wearable wireless system. The PIFA is usually required to fit the shape of a body part such as an arm, head or shoulder, and to be unobtrusive.[20,47,48] The curved radiator and ground plane of the PIFA may affect the impedance and radiation responses.

Properties. A ground plane may not be a conventional PEC plate such as a piece of copper sheet or metal tape. It can also be a modified or artificial ground plane with special properties such as slotted PEC plate and periodic structure.[49,50] Slotted ground planes broaden the impedance bandwidth of compact PIFAs greatly. Recently, some periodic structures such as frequency-selective surfaces (FSSs), photonic band-gap (PBG) or electromagnetic band-gap (EBG) structures, artificial magnetic conductors (AMCs) and high-impedance surfaces have been proposed to improve the performance of antennas or reduce their size. For example, an AMC has been used for low-profile and broadband designs.[51–54]

Variations of PIFAs

Besides PIFAs with rectangular radiators, many variations of PIFAs have been used for specific requirements such as compactness, broad bandwidth, multiple bands and low profiles. Figure 4.10 provides some of the commonly used forms.[16,19,55–58] To keep PIFAs compact and allow multi-band operation, the radiators are usually slotted or notched in some way. The slots or notches change the current paths on the radiators. They also separate the radiator into more than one resonant region for multi-band operations. In practice, the radiators of PIFAs can be of any shape, determined by the installation environment and neighbouring circuits.

4.3.2 PLANAR INVERTED-L ANTENNA

As design examples, two modified PILAs are introduced. One example is a PILA with an L-shaped radiator.[18] The design demonstrates an alternative way of modifying the radiator to achieve a broad bandwidth by notching the radiator.[17,59] The other example modifies the feeding structure to improve the broadband performance of the PILA.[15,60]

An L-shaped PILA

A probe-fed PILA is examined here.[18] This parametric study will provide more information on the effects of the height, the dimensions of the plate and the position of the feed point on

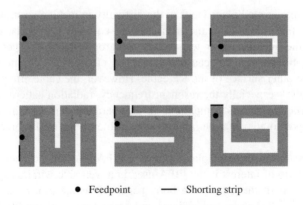

● Feedpoint   — Shorting strip

Figure 4.10 Several modified versions of planar inverted-F antennas.

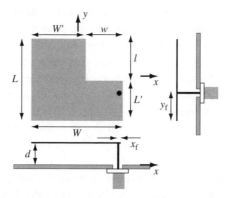

Figure 4.11 L-shaped planar inverted-L antenna. (Reproduced by permission of AGU.[18])

the antenna's impedance characteristics. The concept of the *matching factor* (MF) is used to assess the matching property of a broadband antenna.[61]

Consider a PILA with an L-shaped plate, which is suspended above a ground plane as shown in Figure 4.11. The plate with dimensions W, L, W' and L' is positioned on the x–y plane at a height $z = d$. A coaxial probe excites the plate at location (x_f, y_f) through a 50-Ω SMA connector. Based on simulations and experimental adjustments, the optimized dimensions of the plate antenna for the impedance bandwidth are $W = 50$ mm, $L = 45$ mm, $W' = 30$ mm, $L' = 23$ mm, $d = 7.5$ mm and $(x_f, y_f) = (1$ mm, 16 mm$)$. The plate can also be considered as a radiator with a notch of dimensions l and w.

In the tests, a copper plate with dimensions 310 mm $\times$ 225 mm is used to approximate the infinite ground plane. A foam layer with $\epsilon_r = 1.07$ supports the plate.

The simulated and measured VSWR against frequency are shown in Figure 4.12 for validation. The simulated and measured bandwidths for VSWR $= 2$ are 23.7 % and 24.4 %, respectively. The maximum shift between the simulated and measured frequencies within the bandwidth is about 3 %, which is mainly due to the assumption of a uniform current on the long probe in the EM simulator (Ensemble 5.0).

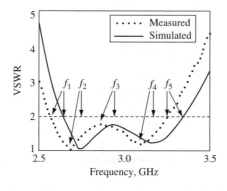

Figure 4.12 Simulated and measured VSWR for the L PILA. (Reproduced by permission of AGU.[18])

Next, a parametric study is carried out by means of the EM simulator. The effect of the spacing d on the input impedance and bandwidth are examined, where the spacing varies from 5 mm to 10 mm. The achieved impedance bandwidths for different spacings are plotted in Figure 4.13. The study shows that the resonant frequencies for two well-matched modes around 2.75 GHz and 3.2 GHz decrease as the spacing increases. The impedance bandwidth is greatest when the spacing is around 8 mm. Beyond that, the bandwidth decreases because the larger spacing leads to a poorer impedance match due to higher inductance. It can be concluded that the frequencies of interest shift down owing to the longer probe when the spacing increases, and bandwidths in the order of 20 % can be achieved when the spacing is between 6 mm and 11 mm.

The matching performance of a broadband antenna is evaluated directly by the MF. The matching factor denotes an average VSWR within the bandwidth for VSWR $= 2$ (usually describing the matching performance of an antenna). The MF is defined by equation (3.1). For example, the measured VSWR for a PILA with two different widths L as illustrated in Figure 4.14, which shows that bandwidths of 22.6 % ($L = 45$ mm) and 23.4 % ($L = 46$ mm) are obtained for VSWR $= 2$. However, it should be noted that, within the bandwidth, the former design has a much better matching condition (MF $= 1.629$) than the latter (MF $= 1.791$), although the former's bandwidth is somewhat smaller than the latter's. This confirms that the observation from Figure 4.14 coincides well with the information derived from the relevant MF. This simple way of evaluating the matching condition of a broadband antenna is useful in deciding design tradeoffs when the basic antenna parameters are considered simultaneously, such as dimensions, operating bandwidth, matching condition and so on. Thus, in the following study, the MF will be used to assess the influences of the antenna parameters on the impedance characteristics.

Figure 4.15 shows the simulated VSWR for different locations (x_f, y_f) of the feed point and the achieved impedance bandwidths with the MF. The MF remains at or below 1.5 for all cases, even though the bandwidth has changed. This suggests that the bandwidth is more sensitive to the location of the feed point than the matching condition.

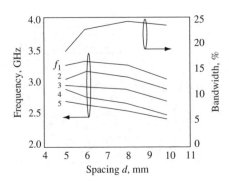

Figure 4.13 Comparison of frequencies and bandwidths for varying d for the PILA. (Reproduced by permission of AGU.[18])

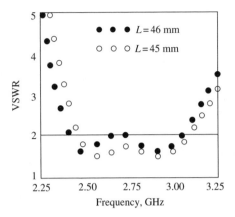

Figure 4.14 Comparison of measured VSWR for varying L for the PILA. (Reproduced by permission of AGU.[18])

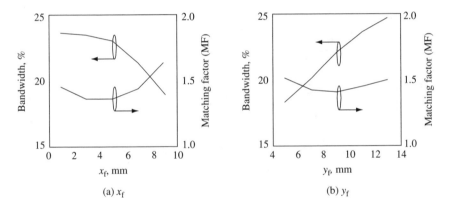

Figure 4.15 Effects of location of the feed point on bandwidths and MF for the PILA. (Reproduced by permission of AGU.[18])

The effects of L and W on the bandwidth and MF are revealed in Figure 4.16. It is not surprising that both bandwidth and MF are sensitive to L and W. The bandwidth and MF increase as L increases but decrease with W.

The effects of the dimensions L' and W' of the rectangular notch are revealed in Figure 4.17. The results suggest that the bandwidth decreases as the dimensions of the notch increase. It is evident that introduction of the rectangular notch leads to an additional well-matched adjacent mode operating at the lower frequency so that the bandwidth increases. For narrower bandwidths, the MF decreases (the matching condition becomes better) because the two adjacent modes approach each other as the dimensions of the notch are smaller.

Simulated electric current distributions on the plate are illustrated in Figure 4.18. It shows the distributions of the imaginary part of the induced currents at two operating frequencies,

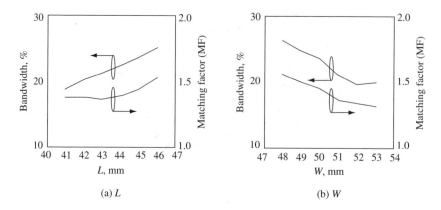

(a) L   (b) W

Figure 4.16 Effects of dimensions L and W of the plate on bandwidths and MF for the PILA. (Reproduced by permission of AGU.[18])

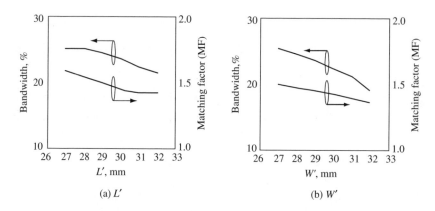

(a) L' (b) W'

Figure 4.17 Effects of dimensions L' and W' of the notch on bandwidths and MF for the antenna. (Reproduced by permission of AGU.[18])

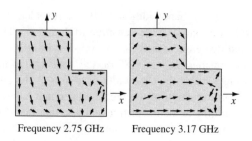

Frequency 2.75 GHz   Frequency 3.17 GHz

Figure 4.18 Distribution of the induced electric currents at the plate of the PILA. (Reproduced by permission of AGU.[18])

2.75 GHz and 3.17 GHz. (The real part is very weak compared with the imaginary part.) This suggests that the plate antenna operates at two different modes with different polarizations. Therefore, a notch cut from the plate antenna can be used to further broaden the bandwidth up to 20 %. This type of antenna, like other PILAs or PIFAs, is suitable for applications not requiring high polarization purity.

A PILA with Vertical Feeding Structure

As previously noted, one way to improve the impedance bandwidth of a PILA is to modify its feeding structure.[13–15] This section introduces a new feeding arrangement applied to a broadband PILA.[13] This design has a radiating element made of a perfectly electrically conducting (PEC) plate parallel to a ground plane at a height of around 0.1 times the center operating wavelength. The feeding structure consists of a pair of PEC strips separated by a thin vertical dielectric slab grounded at its bottom. One of the strips is excited by a coaxial probe at its bottom through an SMA. The other strip is fully grounded at its bottom and open-circuit at its top end, which functions as a vertical ground plane (wall). The suspended PEC plate is perpendicular to the vertical wall and fed by the probe-driven strip at the midpoint of its edge close to the wall, as shown in Figure 4.19.

A thin, square copper plate with sides $L = 70$ mm is selected as a radiator and suspended parallel to the ground plane at height $h = 10$ mm. The feeding structure comprises a thin rectangular dielectric slab (Roger4003, $\epsilon_r = 3.38$) with dimensions 10 mm $(h) \times 50$ mm $(w) \times 32$ mil (t) (about 0.81 mm). The slab is mounted vertically to the ground and grounded at its bottom. A pair of parallel strips are centrally etched on to the two surfaces of the dielectric slab. One of the strips with width $w_g = 50$ mm is grounded at its bottom and

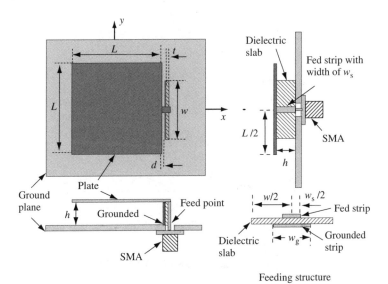

Figure 4.19 Geometry of the PILA with a modified feeding structure. (Reproduced by permission of IEE.[13])

open-circuit at its top end. A 50-Ω coaxial probe with diameter 1.2 mm excites the other strip with width $w_s = 2$ mm at its bottom through an SMA. The probe-driven strip feeds the plate at the midpoint of the plate edge through a narrow strip, which measures 2 mm in width and $d + 32$ mil (dielectric slab thickness, t) in length. The driven plate edge is close to and parallel to the top end of the vertical wall. The term d indicates the feed gap between the plate edge and the top end of the wall. In the tests, a PEC plate with dimensions 310 mm × 330 mm was used to approximate the infinite ground plane.

Figure 4.20(a) shows the achieved bandwidths for VSWR = 2, which reach up to 48 % with the range 2.14–3.5 GHz for $d = 1.0$ mm, and 52 % with the range 2.06–3.52 GHz for $d = 1.5$ mm. The measured input impedance plotted in Figure 4.20(b) demonstrates the achievement of good impedance matching between two adjacent operating modes. Manifestly, the feed gap d has a significant effect on the matching condition or the impedance bandwidth because of the strong electromagnetic coupling between the plate edge and the vertical wall. The plate edge and the vertical wall form the impedance matching network between the plate and the probe. Therefore, adjustments to both the feed gap and wall size

(a) VSWR

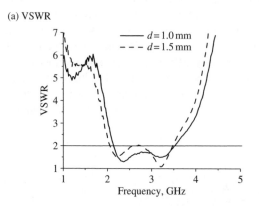

(b) Input impedance

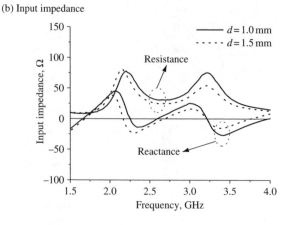

Figure 4.20 Measured VSWR and input impedance of the PILA with a modified feeding structure for varying d. (Reproduced by permission of IEE.[13])

allow good impedance matching across a broader frequency range than with a conventional PILA.

Figure 4.21(a) shows the measured VSWR for various wall widths, namely $w_g = 0, 20$ and 50 mm. All the vertical dielectric slabs measure $h \times w \times t = 10$ mm $\times$ 50 mm $\times$ 32 mil (about 0.81 mm), and the feed gap is maintained at $d = 1.0$ mm. It can be seen that the achieved VSWR $= 2$ impedance bandwidths of 7.3 % (1.84–1.98 GHz) for $w_g = 0$ mm, and 9.9 % (2.11–2.33 GHz) for $w_g = 20$ mm, are much narrower than the bandwidth of 50 % for $w_g = 50$ mm. Evidently, the width of the wall significantly affects the bandwidth because of the strong electromagnetic coupling between the plate and the wall. Therefore, one can readily control the bandwidth by adjusting the feed gap and/or the width of the wall. Additionally, it is also observed from a comparison of the measured VSWR that the resonant frequencies for the three cases are different; that is, 1.91 GHz for $w_g = 0$ mm, 2.23 GHz for $w_g = 20$ mm, and both 2.34 GHz and 3.45 GHz for $w_g = 50$ mm. The wider the wall, the higher is the first resonant frequency. The probe-driven strip radiates less energy into space due to the wider wall. The feeding structure acts as a transmission-line system when the wall becomes much

(a) VSWR

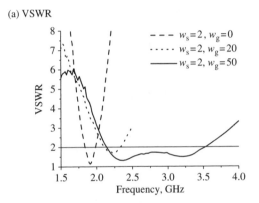

(b) Input impedance

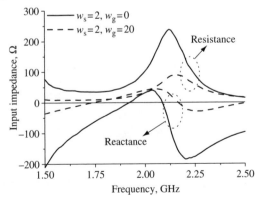

Figure 4.21 Measured VSWR and input impedance of the PILA with a modified feeding structure for varying w_g (dimensions in millimeters). (Reproduced by permission of IEE.[13])

wider than the strip, such as the ratio of $w_g/w_s > 10$. By contrast, it functions as a radiator when the width of the wall approaches zero.

Figure 4.21(b) shows the measured input impedance. As can be seen, by introducing the grounded strip the peak input resistance of 250 Ω decreases to around 70 Ω and the reactance is cancelled out well within the bandwidth. Therefore, the vertical feeding section functions not only as a transmission line but also as an impedance matching network.

The radiation patterns of the PILA were measured in three principal x–z, y–z and x–y planes. The radiation patterns of both electric field components E_θ and E_ϕ were measured at the frequencies 2.3, 2.8 and 3.5 GHz. The results demonstrate that the radiation patterns in each plane are quite similar to those of conventional inverted-F or -L antennas. Figures 4.22 and 4.23 show that the vertical wall does not severely degrade the substantive radiation patterns of the conventional PILAs. Therefore, the antenna has different radiation

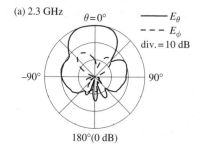

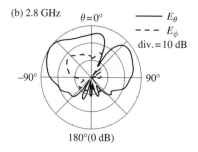

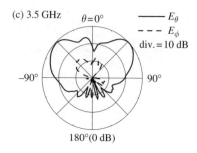

Figure 4.22 Measured radiation patterns at x–z planes. (Reproduced by permission of IEE.[13])

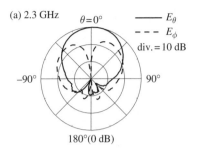

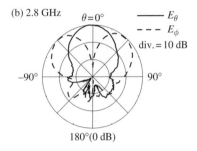

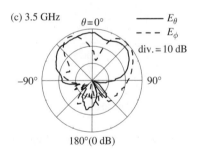

Figure 4.23 Measured radiation patterns at y–z planes. (Reproduced by permission of IEE.[13])

characteristics at different operating frequencies in different modes. Moreover, Figure 4.24 shows that at the higher frequencies of the bandwidth, the radiation patterns in the x–y planes are seriously distorted owing to the impingement of the vertical wall on the radiation. Therefore, the bandwidth for the radiation pattern is narrower than that for the impedance.

In short, this study demonstrates that PILAs with the new feeding structure can provide an impedance bandwidth in the order of 50 % for VSWR = 2, and the radiation patterns are quite similar to those of conventional PIFAs and PILAs. The features of the design suggest that these PILAs have potential applications in land mobile communication systems in unpredictable environments.

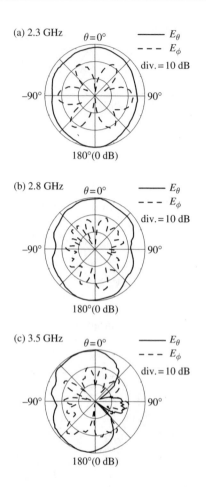

Figure 4.24 Measured radiation patterns at x–y planes. (Reproduced by permission of IEE.[13])

4.4 CASE STUDIES

4.4.1 HANDSET ANTENNAS

As noted earlier, measurements taken with antennas on finite-size ground planes is an important issue in the evaluation of designs. This section describes the effects of the signal cables on operating frequencies, radiation patterns and gain of handset antennas under test (AUTs). With a simple testing scheme, a coaxial signal cable feeds the AUTs through a microstrip transmission line etched on to a printed-circuit board. The influences of different feeding schemes and cable arrangements on a dual-band monopole antenna and embedded PIFA are compared. The distributions of the magnetic fields close to the AUTs, handset chassis and the cable are simulated and measured to reveal the cable-related influences.

In standard antenna measurements, the AUT is put on a large ground plane and connected to a network analyzer through the ground plane by a signal cable. The influence of the signal cable on the measurements is ignored owing to the electrically large ground plane. However, this scheme is not valid for a handset antenna, owing to its electrically small ground plane and portability. The cable used in measurements acts not only as a transmission component but also as a radiator by virtue of the currents traveling on its outer surface. This additional radiation has a significant impact on the antenna performance.

Some methods have been suggested to alleviate this problem. For example, ferrite chokes around the front end of the cable have been used to absorb the energy on the outer surface of the cable.[43] However, the chokes also reduce the total radiated power from the AUT and even change the operating frequency.[45] Another limitation on the use of ferrite chokes is that they typically operate well at frequencies lower than 1 GHz. Alternatively, sleeve-like chokes, such as quarter-wavelength long chokes, were employed to suppress the induced currents flowing on the outer surface of the cable.[44-46] Furthermore, tapered and folded baluns have been designed for measurements on dual-band antennas operating at 900 MHz and 1900 MHz.[62] The performance of a frequency-dependent balun is very responsive to its geometric parameters. Lastly, measurement setups without any cable have been presented for accurate testing,[63,64] but special devices and systems are required.

Investigations into cable-related effects on radiation patterns of AUTs have been implemented numerically. Discussions on the interaction between an antenna, a handset chassis and a feed in the measurements of handset antennas have also been reported. For example, a regular semi-rigid coaxial cable, a sleeve balun and a ferrite choke were selected to weaken the influence of the signal cable at 1900 MHz.[65]

The Testing Scheme and AUTs

Figure 4.25 shows the testing scheme and the coordinate system used. A 50-Ω microstrip transmission-line etched on to a PCB (40 mm × 100 mm × 0.7 mm) connects the AUT to a

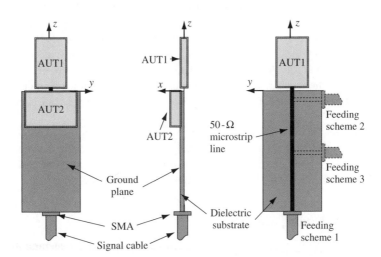

Figure 4.25 The testing scheme. (Reproduced by permission of © IEEE. Z. N. Chen, N. Yang, Y. X. Guo, and M. Y. W. Chia, 'An investigation into measurement of handset antennas,' *IEEE Transactions on Instrumentation and Measurement*, vol. 54, no. 3, pp. 1100–1110, June 2005.)

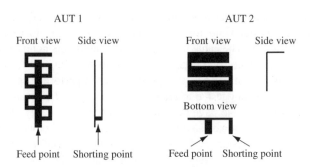

Figure 4.26 Geometry of AUTs. (Reproduced by permission of © IEEE. Z. N. Chen, N. Yang, Y. X. Guo, and M. Y. W. Chia, 'An investigation into measurement of handset antennas,' *IEEE Transactions on Instrumentation and Measurement*, vol. 54, no. 3, pp. 1100–1110, June 2005.)

signal cable via an SMA. The dielectric layer of the PCB is Roger4003 with a dielectric constant of 3.38. One side of the PCB is fully covered with foil and connected (grounded) to the outer surface of the signal cable.

Figure 4.26 displays schematics of two AUTs operating in the GSM900, PCS1900 and ISM2450 bands, which are installed on the PCB. AUT 1 is basically a monopole configuration, namely a combination of a simple monopole and a meandered monopole, which features vertical polarization. AUT 2 is an embedded antenna, namely an S-shaped PIFA, which has the property of dual polarization. The signal cable will be located at three positions on the PCB, namely feeding schemes (FS) 1, 2 and 3 as shown in Figure 4.25.

Measurements of return loss were implemented by an HP8510C network analyzer. The calibration was done at the interface of the SMA output. The signal cable is 1 m in length. Measurements of far-field radiation patterns were conducted in an EM anechoic chamber (4 m × 4 m × 10 m) with a frequency range of 0.2–40 GHz by an Orbit Far-field Antenna System. Measurements of the *H*-fields close to the AUTs, handset chassis and signal cables were executed by a Dosimetric Analysis System (DASY 3) (Schmid and Partner Engineering AG). Figure 4.27 shows the installation diagram for radiation pattern measurements. Three types of cable arrangement (CA) are under consideration. CA 1 is the use of a straight signal cable (vertically positioned). In CA 2, the signal cable is bent into a semicircle of radius 100 mm and vertically installed. CA 3 also uses a straight signal cable but is enclosed by a thick absorber layer.

Also, to validate the measurements, AUT 1 was simulated by a SEMCAD software package based on an FDTD method. In the analyses, an ideal case is considered where an ideal port (source) is employed to excite the transmission line instead of the SMA connected to the signal cable and the AUT 1 directly without any transmission line.

Measurement with Feeding Schemes

The measured results show that the AUTs achieved good impedance matching for GSM900, PCS1900 and ISM2450 bands at the same time. Within all the bands, the effects of the testing schemes on the performance of the AUTs, such as operating frequency, radiation pattern, gain and even the distribution of the *H*-field near to the AUTs, handset chassis and signal cables were investigated.

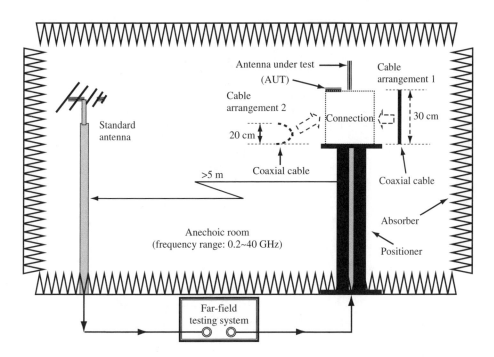

Figure 4.27 Radiation pattern measurement set up. (Reproduced by permission of © IEEE. Z. N. Chen, N. Yang, Y. X. Guo, and M. Y. W. Chia, 'An investigation into measurement of handset antennas,' *IEEE Transactions on Instrumentation and Measurement*, vol. 54, no. 3, pp. 1100–1110, June 2005.)

Operating frequency. Figure 4.28 shows the measured and simulated return losses ($|S_{11}|$) of AUT 1 and 2 with feeding schemes 1, 2 and 3. Here, the operating frequency is defined as the frequency at which the return loss reaches a minimum. Several important points can be drawn.

First, in both GSM and PCS bands, the measured operating frequencies of AUT 1 with all the feeding schemes are in good agreement with the simulated ideal ones. Compared to the calculated frequencies, the measured operating frequency shift varies from 0 MHz to 100 MHz. In particular, the differences between the ideal and actual cases with FS 3 are much less than 20 MHz; that is, 2.4 %.

Second, the operating frequencies for FS 1 are higher than those for FS 2 and 3, even for the ideal case. The simulation shows that the current distributions on the ground plane for FS1 have been changed. The out-of-phase currents along the transmission line on the PCB cancel out the part of the radiation from the ground plane so that the effective length of the ground plane as a radiator is reduced.

Third, the operating frequencies for FS 2 are lower than the simulated ones. This is because the location of the signal cable is quite close to the feed points of the AUTs so that large currents flow onto the outer surface of the signal cable and increases the effective size of the radiator. This point can also be observed from the current distributions shown later.

Lastly, the effects of the signal cable on the operating frequencies of AUT 2 with the different feeding schemes are relatively slight as against AUT 1, since more current is concentrated around AUT 2.

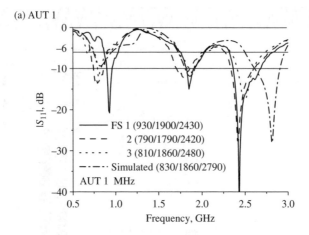

(a) AUT 1

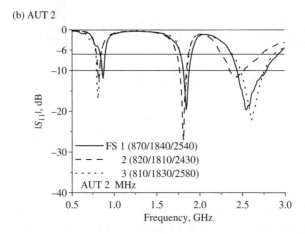

(b) AUT 2

Figure 4.28 Simulated and measured return loss for AUT 1 and 2. (Reproduced by permission of © IEEE. Z. N. Chen, N. Yang, Y. X. Guo, and M. Y. W. Chia, 'An investigation into measurement of handset antennas,' *IEEE Transactions on Instrumentation and Measurement*, vol. 54, no. 3, pp. 1100–1110, June 2005.)

Therefore, during tests, the influence of the feeding scheme on the operating frequency should be taken into account carefully because there could be pronounced frequency shifts of as much as 10 %.[67,22] In the lower bands (GSM/PCS), the signal cable has a more significant impact on the operating frequencies of a monopole than of an embedded antenna.

Radiation patterns. The shapes of radiation patterns and the achieved maximum gain in three primary planes were investigated.

First, AUT 1 was measured with the three types of CAs and with FS 1. Due to the symmetrical structure, AUT 1 has almost the same radiation patterns in ϕ-cuts. Thus, Figure 4.29 compares the simulated and measured E_θ-component radiation patterns in the GSM900 band in three principal planes. In the x–y (horizontal) plane, the radiation pattern for CA 2 has a 2.5-dB dip around the direction $\phi = 260°$. Furthermore, the radiation patterns for both CA 1 and 3 are almost omnidirectional. In the x–z and y–z planes (vertical) planes,

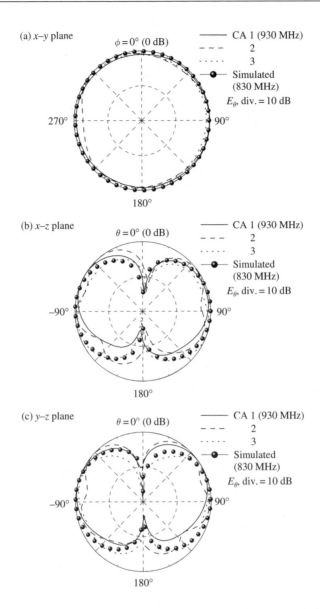

Figure 4.29 Comparison between simulated and measured E_θ-component radiation patterns in the GSM900 band. (Reproduced by permission of © IEEE. Z. N. Chen, N. Yang, Y. X. Guo, and M. Y. W. Chia, 'An investigation into measurement of handset antennas,' *IEEE Transactions on Instrumentation and Measurement*, vol. 54, no. 3, pp. 1100–1110, June 2005.)

the radiation patterns for the CA 2 (with a bent signal cable) are undulating with three additional lobes because of the bent signal cable.

The simulated and measured radiation patterns in the PCS band are plotted in Figure 4.30. In the x–y plane, the asymmetry of the measured radiation patterns for all the CAs ranges from 2 dB to 4 dB but no dip appears. In contrast to the cases in the GSM band, all the

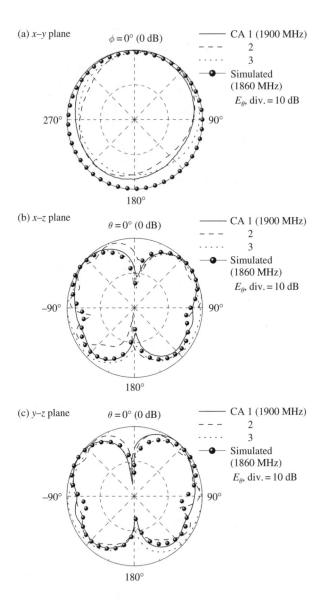

(a) x–y plane $\phi = 0°$ (0 dB)

——— CA 1 (1900 MHz)
– – –   2
· · · · ·   3
—●—   Simulated (1860 MHz)
E_θ, div. = 10 dB

270°   90°

180°

(b) x–z plane $\theta = 0°$ (0 dB)

——— CA 1 (1900 MHz)
– – –   2
· · · · ·   3
—●—   Simulated (1860 MHz)
E_θ, div. = 10 dB

–90°   90°

180°

(c) y–z plane $\theta = 0°$ (0 dB)

——— CA 1 (1900 MHz)
– – –   2
· · · · ·   3
—●—   Simulated (1860 MHz)
E_θ, div. = 10 dB

–90°   90°

180°

Figure 4.30 Comparison between simulated and measured E_θ-component radiation patterns in the GSM1900 band. (Reproduced by permission of © IEEE. Z. N. Chen, N. Yang, Y. X. Guo, and M. Y. W. Chia, 'An investigation into measurement of handset antennas,' *IEEE Transactions on Instrumentation and Measurement*, vol. 54, no. 3, pp. 1100–1110, June 2005.)

simulated and measured radiation patterns in the x–z planes are undulating with at least two additional lobes. The additional lobes result from the electrically large ground plane. However, for the case with CA 2, the distorted radiation patterns with at least six additional side lobes are caused mainly by the undesirable radiation from the signal cable. The radiation patterns were also examined in the ISM band (not shown here). Owing to the impact of the large PCB and the signal cable, the radiation patterns are severely deteriorated, which is similar to the case in the PCS band.

Next, AUT 2 was measured with CA 1 and 2 and with FS 1. Owing to the complicated configuration of AUT 2, the radiation of both E_θ and E_ϕ components must be considered. For brevity, Figure 4.31 compares the radiation patterns only in the x–y and y–z planes, although the patterns in the x–z and y–z planes are not completely the same. The radiation patterns are normalized by the maxima of the E_θ and E_ϕ components. The measurements show that the radiation levels of the E_ϕ components are nearly as high as those of the E_θ components. Moreover, the bent cable substantially affects the radiation patterns of AUT 2. In particular, in the GSM band, the impingement caused by the bent cable on the radiation patterns is more severe than that in the PCS band. The same conclusion has been made in the discussion on AUT 1, which can be further verified by the near-field distribution of H-fields (electric currents) of the AUTs.

Gain. The achieved gain in the direction of the maximum E_θ component in each of the three principal planes should be one of the important parameters in handset antenna design. Thus, the maximum gain for the testing schemes with the different CAs and FSs in the three principal planes are shown in Table 4.1.

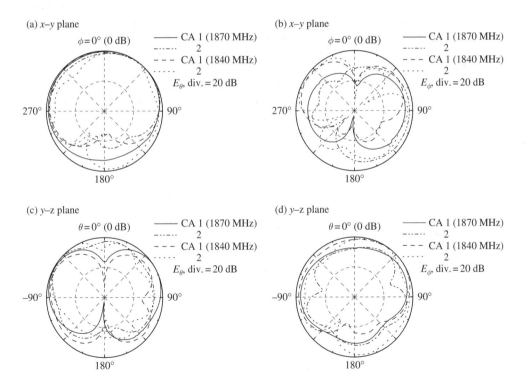

Figure 4.31 Comparison between the radiation patterns in the x–y and y–z planes. (Reproduced by permission of © IEEE. Z. N. Chen, N. Yang, Y. X. Guo, and M. Y. W. Chia, 'An investigation into measurement of handset antennas,' *IEEE Transactions on Instrumentation and Measurement*, vol. 54, no. 3, pp. 1100–1110, June 2005.)

Table 4.1 Comparison of the effects of the testing schemes on gain.

FS[a]	CA[b]	Gain, (dBi) and f (MHz) for AUT 1			Gain, (dBi) and f (MHz) for AUT 2		
		GSM900	GSM1900	ISM2450	GSM900	GSM1900	ISM2450
1	0	1.3/830	4.3/1860	—/—	—/—	—/—	—/—
1	1	1.0/930	3.8/1900	5.2/2430	0.9/87	3.9/184	5.0/2540
1	2	1.3/930	3.8/1900	3.6/2430	1.1/870	5.0/1840	6.0/2540
1	3	−0.6/930	3.2/1900	4.2/2430	—/—	—/—	—/—
2	0	1.1/860	4.1/1880	—	—/—	—/—	—/—
2	1	1.2/790	3.5/1790	5.7/2420	1.0/820	2.5/1810	3.3/2430
3	0	1.2/830	4.3/1860	—	—/—	—/—	—/—
3	1	1.3/810	3.8/1860	4.8/2480	1.2/810	3.1/1830	5.8/2580

[a] FS, feeding scheme.
[b] CA, cable arrangement: 0 = no cable (simulation); 1 = a straight cable; 2 = a bent cable; 3 = a straight cable with an absorber layer.

The measured gain is slightly less than the simulated ones because part of the energy travels on to the cable which is considered as a loss. The measured cross-polarized radiation levels are usually higher than the simulated ones due to the measurement environment. Therefore, the CAs significantly affect the measured gain, especially at the lower frequencies. In particular, by using CA 3, the radiation patterns can be improved as shown in Figure 4.32(a). However, it should be noted that, compared to CA 1 and 2, the measured maximum gain for CA 3 with FS 1 is reduced by about 1.6 dBi at 930 MHz and 0.6 dBi at 1900 MHz in the principal planes because of the absorption of the energy by the absorber layer. Figure 4.32 displays the radiation patterns for the E_θ and E_ϕ components of AUT 1 with CA 3 and FS 1 in the y-z plane. It is seen that the radiation of the E_ϕ components, namely the cross-polarized components, increases to a great extent and is directed to the bottom of the handset.

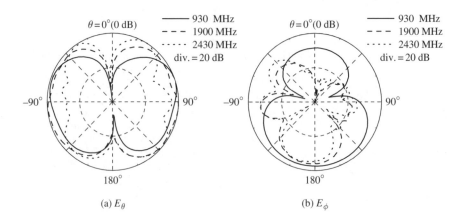

(a) E_θ (b) E_ϕ

Figure 4.32 Measured radiation patterns in y–z planes for AUI 1/u. (Reproduced by permission of © IEEE. Z. N. Chen, N. Yang, Y. X. Guo, and M. Y. W. Chia, 'An investigation into measurement of handset antennas,' *IEEE Transactions on Instrumentation and Measurement*, vol. 54, no. 3, pp. 1100–1110, June 2005.)

Additional tests. Based on the simulated and measured results, the effects of the position of the cable feed on the radiation properties of the AUTs with CA 1 were also investigated as follows.

For AUT 1, Figures 4.33(a) and (b) show that, in the x–y (horizontal) plane, the radiation patterns for AUT 1 are distorted in the PCS band. The maximum differences between the simulated and measured results for FS 1, 2 and 3 are 4 dB, 17 dB and 8 dB, respectively. The radiation patterns in the GSM band remain nearly omnidirectional.

Figures 4.33(c) and (d) exhibit the radiation patterns for AUT 1 in the x–z (vertical) plane. In both GSM and PCS bands, the measured radiation patterns for FS 2 and 3 have very good agreement with the simulations. The comparison demonstrates that, for AUT 1, the FS 2 and 3 slightly impinge on the vertical radiation patterns but they severely affect the horizontal radiation patterns at the higher frequencies. In addition, a squint of less than 10° appears in some radiation patterns in the x–z planes, owing to the asymmetrical positions of the signal cables connected to AUT 1 with feeding schemes 2 and 3.

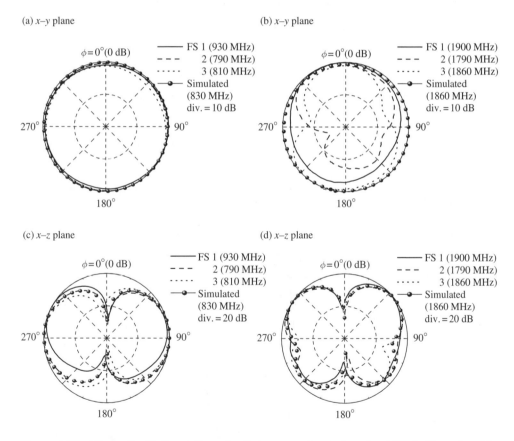

Figure 4.33 Measured radiation patterns for E_θ in x–y and x–z plane for AUT 1. (Reproduced by permission of © IEEE. Z. N. Chen, N. Yang, Y. X. Guo, and M. Y. W. Chia, 'An investigation into measurement of handset antennas,' *IEEE Transactions on Instrumentation and Measurement*, vol. 54, no. 3, pp. 1100–1110, June 2005.)

In contrast to AUT 1, the impact of the signal cable with the different feeding schemes for AUT 2 is more complicated and severe owing to its polarization properties. For brevity, Figures 4.34(a) and (b) show only the radiation patterns of the E_θ components in the x–y and x–z planes. The comparison between the different feeding schemes shows that AUT 2 with FS 1 has severely distorted radiation patterns in both planes. However, the radiation patterns for FS 2 and 3 are almost the same.

Example of Testing a Commercial Handset Antenna

As an example, a commercial antenna for the Nokia 8210 handset operating at 900 MHz and 1800 MHz was measured using the testing scheme with FS 3 and CA 1. The measured

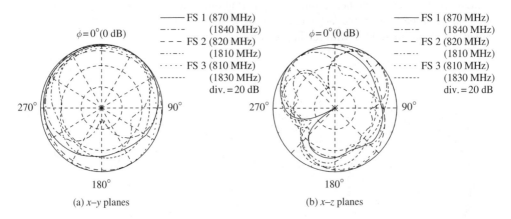

(a) x–y planes   (b) x–z planes

Figure 4.34 Measured radiation patterns for E_θ in x–y and x–z planes for AUT 2. (Reproduced by permission of © IEEE. Z. N. Chen, N. Yang, Y. X. Guo, and M. Y. W. Chia, 'An investigation into measurement of handset antennas,' *IEEE Transactions on Instrumentation and Measurement*, vol. 54, no. 3, pp. 1100–1110, June 2005.)

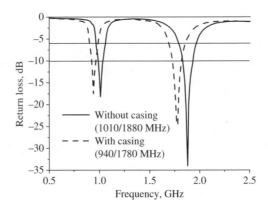

Figure 4.35 Measured return loss for the antenna in a Nokia 8210 handset. (Reproduced by permission of © IEEE. Z. N. Chen, N. Yang, Y. X. Guo, and M. Y. W. Chia, 'An investigation into measurement of handset antennas,' *IEEE Transactions on Instrumentation and Measurement*, vol. 54, no. 3, pp. 1100–1110, June 2005.)

return loss is plotted in Figure 4.35. The operating frequencies are 1010 MHz/1880 MHz and 940 MHz/1780 MHz for the cases without and with the plastic casing, respectively. The central frequency with the plastic casing has shifted down by about 5 %. The measured radiation patterns are depicted in Figure 4.36. The achieved maximum gain in three principal planes without the casing is 1.6 dBi at 1010 MHz and 4.3 dBi at 1880 MHz. All the measured

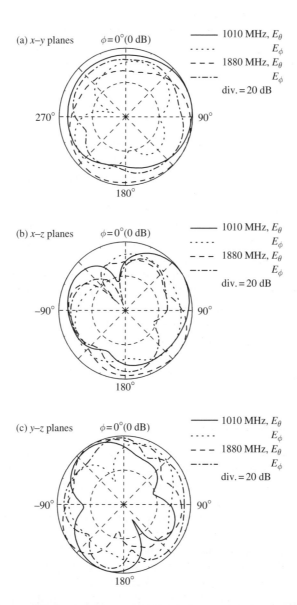

Figure 4.36 Measured radiation patterns for the antenna in a Nokia 8210 handset. (Reproduced by permission of © IEEE. Z. N. Chen, N. Yang, Y. X. Guo, and M. Y. W. Chia, 'An investigation into measurement of handset antennas,' *IEEE Transactions on Instrumentation and Measurement*, vol. 54, no. 3, pp. 1100–1110, June 2005.)

parameters meet the actual operation requirements well. Therefore, the testing scheme with FS 3 and CA 1 is an acceptable option for measurements on handset antennas.

Statistical Analysis

Statistical analysis of radiation patterns can be carried out to further evaluate the cable-related effects. Using the term *mean pattern difference* (δ), the average effects of the CA and the FS on the radiation patterns in different bands and planes are assessed statistically. The mean pattern difference (MPD) in a specific plane is defined as follows:

$$\delta = \frac{\sum_{n=1}^{N} |p_n - p_{n0}|}{N}, \text{dB} \tag{4.1}$$

where N stands for the total number of points under consideration in the specific plane, p_n is the value of the measured electric field at the nth point under consideration, and p_{n0} is the reference value of the electric field at the nth point, which in this case is the simulated value.

Figure 4.37(a) shows the statistical data for the different CAs in GSM900 (Band 1) and PCS (Band 2). In the x–y plane, the value of δ is lower than or equal to those in the x–z and y–z planes. The MPDs at lower frequencies are about 1 dB lower than those at higher frequencies. In all the planes and frequency bands, the cases with CA 2 have the maximum MPDs, and the effects of the cables on the radiation patterns in the x–z and y–z planes are different. Therefore, the use of CA 2 should be avoided in tests. Similarly, Figure 4.37(b) displays the statistical data for the different FSs also in GSM900 (Band 1) and PCS (Band 2). The MPDs for FS 2 are lowest (less than 1.2 dB) in Band 1. In Band 2, the MPDs for FS 1 and 3 range from 1.85 dB to 2.61 dB. However, the large MPD of 5.26 dB can be observed in the x–y plane for FS 2 in Band 2. The effect of the signal cable on the radiation patterns in the x–z and y–z planes are almost the same in each band.

Distributions of Electric Currents

The total H-fields close to the AUTs, the PCBs and the signal cables are simulated and measured in the three frequency bands of interest to further reveal the cable-related effects for different types of AUTs in different bands and with different FSs.

The distributions of the magnitude of the electric currents on perfectly electrically conducting (PEC) objects are directly proportional to the distribution of the H-field close to the PEC objects. Thus, the distribution of the H-field well describes the distribution of the induced currents on the objects under discussion.

In the tests, a small 3-dimensional H-field sensor was used, controlled by a fully automatic robot. The AUTs were placed horizontally. The distance h from the sensor to each AUT was varied from 5 mm to 15 mm. For AUT 1, the PCB and the signal cable were horizontally located in the same plane, but AUT 2 was put horizontally at a height of 10 mm above the PCB and the signal cable, as shown in Figure 4.38. As an example, Figure 4.39 displays some measured results for distance $h = 5$ mm.

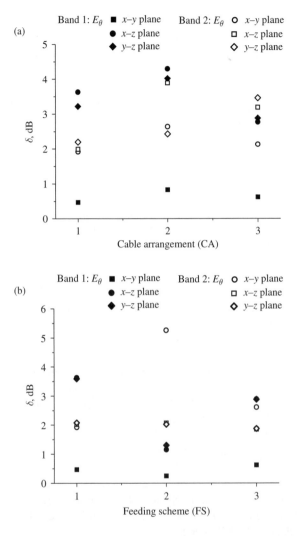

Figure 4.37 Mean pattern difference: (a) for various cable arrangements, and (b) for various feeding schemes. (Reproduced by permission of © IEEE. Z. N. Chen, N. Yang, Y. X. Guo, and M. Y. W. Chia, 'An investigation into measurement of handset antennas,' *IEEE Transactions on Instrumentation and Measurement*, vol. 54, no. 3, pp. 1100–1110, June 2005.)

Figures 4.39(a) and (b) show the current distributions measured at 930, 1900 and 2430 MHz for AUT 1 with FS 1, and at 850, 1840 and 2450 MHz for AUT 2 with FS 1. Figure 4.39(a) shows that the currents on the signal cable connected to AUT 1 become weaker at the higher testing frequencies because the size of the handset (AUT 1 plus the PCB) becomes gradually electrically larger than the half-wavelength with increasing operating frequency. The ratios of maximum *H*-field intensities on the signal cable for AUT 1 are 35 % at 930 MHz, 15 % at 1900 MHz and 10 % at 2430 MHz, because more and more energy is concentrated around AUT 1 with increasing operating frequency.

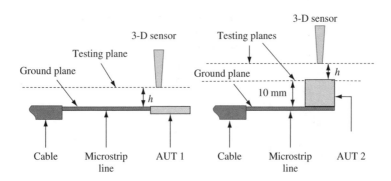

Figure 4.38 Measurement scheme for H-field above the AUTs and ground planes. (Reproduced by permission of © IEEE. Z. N. Chen, N. Yang, Y. X. Guo, and M. Y. W. Chia, 'An investigation into measurement of handset antennas,' *IEEE Transactions on Instrumentation and Measurement*, vol. 54, no. 3, pp. 1100–1110, June 2005.)

This observation clearly reveals the reason why the signal cable has a more significant influence on the performance of AUT 1 at the lower frequencies than at the higher frequencies. It is also evident that the undesired radiation from the signal cable mainly causes the additional lobes in the radiation patterns at the lower frequencies. At the higher frequencies, the radiation from the ground plane primarily undulates the radiation patterns although the unwanted radiation from the signal cable contributes to the distortion as well. However, Figure 4.39(b) suggests that all the ratios of maximum H-field intensities on the signal cable for AUT 2 are nearly 10 % at all the operating frequencies, although the distributions around AUT 2 have changed. These ratios are much lower that those for AUT 1. Comparison of the results for both AUTs suggests that the use of AUT 2 reduces the currents flowing on the outer surface of the signal cable to a great extent; much more energy is concentrated around AUT 2 than AUT 1.

Figure 4.39(a) and Figures 4.40(a) and (b) compare the H-field distributions for AUT 1 with FS 1, 2 and 3 in both GSM and PCS bands. It can be seen that, similar to the case with FS 1, the effect of the signal cable on the distributions for FS 2 and 3 are much more severe in the GSM band than in the PCS band.

Figure 4.39(a) and Figures 4.40(a) and (b) also show that for the same band, namely the GSM *or* PCS band, FS 2 affects the distributions around the antennas more significantly than FS 1 and 3. This results in the variation of the operating frequencies as discussed above. In Figures 4.39 and 4.40 show that, owing to the asymmetrical current distributions, the distortion of radiation patterns in the x–y planes for FS 2 and 3 is much more than that for FS 1 (as shown in Figures 4.33(a) and (b)), although the strong currents appear at the outer surface of the signal cable in FS 1. Furthermore, the weak current distribution on the outer surface of the signal cable for the FS 2 confirms that the radiation patterns for FS 2 are in very good agreement with those of the ideal case, as shown in Figures 4.33(c) and (d).

Therefore, feeding scheme 1 has a slight impact on the horizontal radiation patterns and the operating frequencies, while feeding scheme 2 has a significant influence on the operating frequency and the horizontal radiation patterns but a slight impingement on the vertical radiation patterns.

(a) AUT 1 with FS 1

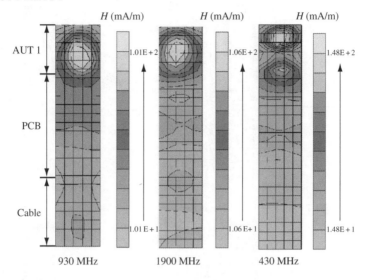

930 MHz   1900 MHz   430 MHz

(b) AUT 2 with FS 1

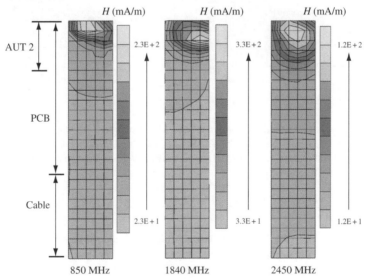

850 MHz   1840 MHz   2450 MHz

Figure 4.39 Measured H-field distributions above AUT 1 and 2 with feeding scheme 1. (Reproduced by permission of © IEEE. Z. N. Chen, N. Yang, Y. X. Guo, and M. Y. W. Chia, 'An investigation into measurement of handset antennas,' *IEEE Transactions on Instrumentation and Measurement*, vol. 54, no. 3, pp. 1100–1110, June 2005.)

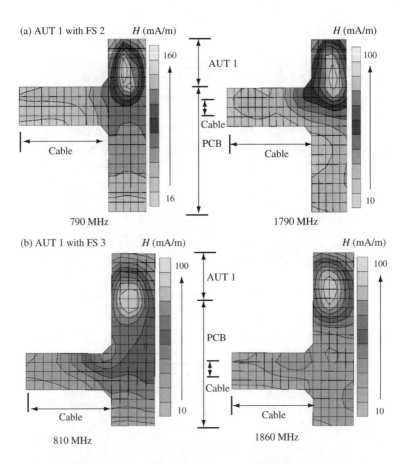

Figure 4.40 Measured H-field distributions above AUT 1 with feeding schemes 2 and 3. (Reproduced by permission of © IEEE. Z. N. Chen, N. Yang, Y. X. Guo, and M. Y. W. Chia, 'An investigation into measurement of handset antennas,' *IEEE Transactions on Instrumentation and Measurement*, vol. 54, no. 3, pp. 1100–1110, June 2005.)

Conclusions

The information derived from of simulations and measurements suggests that the measurement of the impedance and radiation performances of handset antennas is sensitive to the testing scheme. With a testing scheme using a microstrip transmission line, the impact of the signal cable can be reduced if the cable arrangement and the feeding scheme are properly chosen, because the feeding currents flowing on the outer surface of the signal cable to some extent become weak due to the radiation from the ground plane. Four important points have been observed from the measurements of handset antennas using the simple testing method:

- The most significant influence of the signal cable on the performance of a handset antenna occurs at low frequencies, such as in the 900-MHz bands.
- The signal cable affects the performance of a monopole-like antenna more than those of PIFAs.

- The straight cable should be located far from the AUT, which reduces the disturbance of the signal cable on the current distributions near the AUT, so that the measured radiation performance can approach the actual performance.
- The RF absorber enclosing the signal cable can be used to eliminate additional lobes from the radiation patterns to a great extent.

All the conclusions are applicable to the measurements of antennas used in other wireless devices with small system ground planes.

4.4.2 LAPTOP COMPUTER ANTENNAS

The antenna for a cellular phone is often installed on the top of the handset. The radiating element of the embedded PIFA is installed above and parallel to a finite-size ground plane. The volume of the antenna is critical for realizing good performance, especially at low frequency bands. In contrast, antennas embedded into laptop computers are usually installed in the frame of the display (LCD), as showed in Figure 4.41. The cover of the display is made of lossy metallic or plastic material. The antenna being positioned in such a lossy environment, its gain becomes the most important parameter.

The main constraints in antenna design include the thickness and the height of the antenna. Typically the thickness is less than 2 mm, and the height is between 10 mm and 12 mm. The LCD with a thickness of about 5 mm, the cover rim, and the cover itself significantly block the radiation from the embedded antennas and reduce the gain. To alleviate the blockage by the LCD, the antenna is designed with a 5-mm high vertical ground plane. Therefore, the height of the antenna is allowed to be just 5–7 mm in practical applications.

Recently, several wireless communication standards have been established based on use of the 2.45-GHz ISM band. For instance, many of today's laptop computers have incorporated Bluetooth technology as a cable replacement to communicate with other portable and/or fixed electronic devices. By means of IEEE 802.11b technology, WLAN devices can provide data rates up to 11 Mbps. For much higher data rates, WLAN devices based on IEEE 802.11a technology operating in the 5.15/5.85-GHz band are required; such devices have data rates

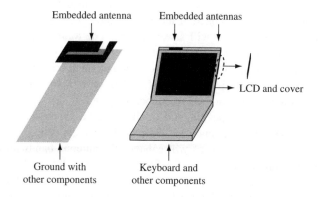

Embedded antenna   Embedded antennas

LCD and cover

Ground with   Keyboard and
other components   other components

Figure 4.41 Comparison of antennas for a handset and a laptop computer.

up to 54 Mbps. The 802.11a devices with proposed channel binding techniques are expected to extend the data rate up to 108 Mbps. Furthermore, 802.11a devices operating in the 5.15/5.85-GHz band also have less interference issues than are faced in the 2.45-GHz ISM band. Future WLAN devices will combine IEEE 802.11a/11b/11g together. In fact, some WLAN devices already have this triple combination. As a result, the demand for an antenna operating at both bands is increasing. For worldwide applications, an antenna covering the 5.15/5.85-GHz range is needed. Many solutions have been used for the multiband applications to cover the 2.45-GHz band, the 5.15/5.85-GHz band, GPS bands, the bands for cellular phones, and even the promising UWB bands. The existing broadband or multiband antennas with one feed point are basically modified PIFAs, where the planar radiating element is co-planar with the vertical ground plane and perpendicular to the cover.[68-71]

A multiband PIFA is here exemplified that can be easily integrated into laptop computers or other portable devices. The design also features cost effectiveness and provides reliable performance. Figure 4.42 shows a 3-dimensional view of the multiband antenna structure. The antenna consists of three parts, namely a PIFA element, a coupled L-shaped element inside the inverted-F element, and an L-shaped branch element connected to the inverted-F element. Since the multiband antenna is an extension of the PIFA to provide two additional frequency bands (middle and high bands due to the branch and coupled elements), it essentially behaves as a PIFA at the low band.

As shown in Figure 4.42(a), the multiband antenna is stamped from thin sheet metal. Figure 4.42(b) shows a sketch of the antenna with geometric parameters. Depending on applications, the antenna can be configured to operate either as a dual-band antenna or a tri-band antenna. For the dual-band antenna, the branch and coupled elements produce two resonances in the high band. The antenna is basically a PIFA at the low band. So, the antenna tuning for the low band is basically the same as the tuning for the PIFA. The lowest resonant frequency f_1 is adjusted by the length $IH + IL$. Changing the height IH will change the frequency f_1. Generally, increasing the height IH will enhance the impedance bandwidth. Adjusting the length IG will change the antenna input impedance at f_1. Increasing IG will increase the antenna input impedance, and vice versa. Adjusting IG will also affect f_1, but its effect is less significant than those of IH and IL. The branch elements can be in different shapes and in different locations either along the elements with the lengths IL and IG or even along the feed element.

However, the resonant frequency f_2 ($f_1 < f_2$) is determined primarily by the branched element length $BH + BL$. The connection location to the PIFA determines the impedance around f_2. The connection location will also affect f_2. The total length ($CH + CL$) of the coupled element primarily determines the resonant frequency f_2 ($f_2 < f_3$). The impedance at f_2 is controlled by the coupling between the coupled element and the inverted-F element. The coupling will be strong if the distances $IH - CH$ or CD are reduced. For the dual-band WLAN applications, two resonant frequencies f_2 and f_3 are used to cover the band from 5.15 GHz to 5.85 GHz. Depending on applications, it is possible that the coupled element produces the resonant frequency f_2, while the branch element produces the resonant frequency f_3. Widening the metal strips will enhance the antenna bandwidth in both bands. Adjusting the resonant frequency at the 5-GHz band has very little effect on the resonant frequency at the 2.45-GHz band. So, in practical applications, one should tune the inverted-F portion first for the low band, then adjust the branch element and coupled element for the second and third resonant frequencies.

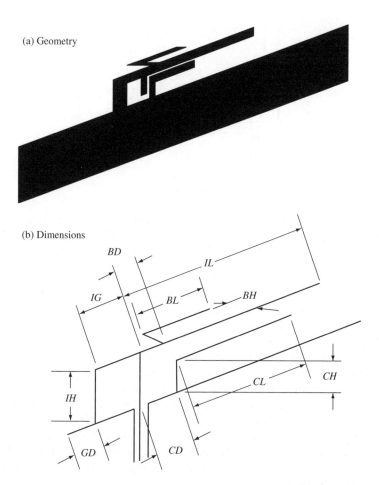

Figure 4.42 Geometry and dimensions of a PIFA for a laptop computer. (Reproduced by permission of © IEEE. D. Liu and B. Gaucher, 'A new multiband antenna for WLAN/cellular applications,' IEEE Vehicular Technology Conference, vol.1, pp. 243–246, September 2004.)

As an example of the design, the main dimensions of the antenna are provided as follows. The rectangular ground plane measures 70 mm × 10 mm × 0.2 mm. The main PIFA has a 2-mm wide horizontal element and a 4-mm wide vertical element. The L-shaped branch portion is 1 mm wide, while the coupled L-shaped portion has a 2-mm wide vertical element and a 1-mm wide horizontal element. The horizontal element is fed by a 1-mm wide strip through a 50-Ω RF cable. Other location parameters include $IH = 6$ mm, $IG = 7$ mm, $CD = 1$ mm, $CH = 3.5$ mm, $CL = 8$ mm, $BD = 4$ mm, $BH = 5$ mm, $BL = 30$ mm and $GD = 20$ mm. Figure 4.43 plots the VSWR for the design with these parameters, which covers both the 2.45-GHz (2.4–2.5 GHz) and the 5-GHz (5.15–5.85 GHz) bands with enough margins. For the 5-GHz band, two resonances exist due to two resonating elements: the branch element and the coupled element. The radiation patterns on the horizontal plane for 2.45-GHz and 5-GHz bands are nearly omnidirectional. For practical purposes, the radiation patterns change

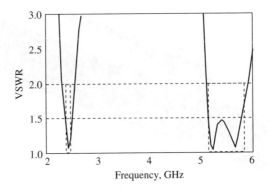

Figure 4.43 VSWR of an embedded PIFA for a laptop computer. (Reproduced by permission of © IEEE. D. Liu and B. Gaucher, 'A new multiband antenna for WLAN/cellular applications,' IEEE Vehicular Technology Conference, vol. 1, pp. 243–246, September 2004.)

very little through the 5-GHz band. The antenna gain is typically 3–4 dBi in both bands. The elevation-plane radiation patterns in both bands and angles of $\phi = 0°$ and 90° do not have deep nulls and are almost omnidirectional.

However, this design was based on a free space environment. For actual applications such as integrated WLAN antennas for laptop computers, the ground plane will overlap with laptop metal display support structures as mentioned previously. So, the actual ground plane can be much smaller. The antenna can be mounted on laptops either perpendicular or parallel to the laptop display surface. In either case, the slot of the inverted-F element has to be clear from any metal structure, otherwise the antenna efficiency will be poor. The key point here is that the antenna must be considered as part of the actual laptop structure and tuned to match those conditions. The effects of the actual environment on the radiation and impedance performance of antennas that are embedded into lossy laptop displays is discussed in the next chapter.

REFERENCES

[1] R. W. P. King, J. C. W. Harrison and J. D. H. Denton, 'Transmission-line missile antennas,' *IRE Transactions on Antennas and Propagation*, vol. 8, no. 1, pp. 29–33, 1960.

[2] R. W. P. King and C. W. Harrison, 'The inverted L-antenna: current and impedance,' Sandia Corp., Albuquerque, NM, technical memo., pp. 353–359, 1949.

[3] K. Kobayashi, S. Nishiki, T. Taga and A. Sasaki, 'Detachable mobile radio units for 800MHz land mobile radio system,' *34th IEEE Vehicular Technology Conference*, vol. 1, pp. 21–23, 1984.

[4] T. Taga and K. Tsunekawa, 'Performance analysis of a built-in planar inverted-F antenna for 800MHz band portable radio units,' *IEEE Journal of Selected Areas of Communications*, vol. 5, no. 5, pp. 921–929, 1987.

[5] K. Hirasawa and M. Haneishi (Ed.), *Analysis, Design, and Measurement of Small and Low-profile Antennas*. Boston, MA: Artech House, 1991.

[6] J. Rasinger, A. L. Scholtz, W. Pichler and E. Bonek, 'Influence of surface waves upon efficiency and mutual coupling between rectangular microstrip antennas,' *IEEE International Symposium on Antennas and Propagation*, vol. 2, pp. 660–663, June 1990.

[7] L. L. Rauth, J. S. McLean, K. B. Dorner, J. R. Casey and G. E. Crook, 'Broadband low-profile antenna for portable data terminal,' *IEEE International Symposium on Antennas and Propagation*, vol. 1, pp. 438–441, June 1997.

[8] H. Nakano, N. Ikeda, Y. Wu, R. Suzuki, H. Mimaki and J. Yamauchi, 'Realization of dual-frequency and wide-band VSWR performances using normal-mode helical and inverted-F antennas,' *IEEE Transactions on Antennas and Propagation*, vol. 46, no. 6, pp. 788–793, 1998.

[9] P. Song, P. S. Hall, H. Ghafouri-Shiraz and D. Wake, 'Triple-band planar inverted-F antennas for handheld devices,' *Electronics Letters*, vol. 36. no. 2, pp. 112–114, 2000.

[10] T. Okuno and K. Hirasawa, 'A semicircular planar inverted-F antenna with a parasitic element,' *8th International Conference on Communication Systems*, vol. 2, pp. 1170–1173, November 2002.

[11] R. L. Li, G. DeJean, M. M. Tentzeris and J. Laskar, 'Developement and analysis of a folded shorted-patch antenna with reduced size,' *IEEE Transactions on Antennas and Propagation*, vol. 52, no. 2, pp. 555–562, 2004.

[12] K. Oh and K. Hirasawa, 'A dual-band inverted-L-folded antenna with a parasitic wire,' *IEEE International Symposium on Antennas and Propagation*, vol. 3, pp. 3131–3134, June 2004.

[13] Z. N. Chen and M. Y. W. Chia, 'Broadband planar inverted-L antennas,' *IEE Proceeedings: Microwave, Antennas and Propagation*, vol. 148, no. 5, pp. 339–342, 2001.

[14] A. Kishk, K. F. Lee, W. C. Mok and K. M. Luk, 'A wide-band small size microstrip antenna proximately coupled to a hook-shaped probe,' *IEEE Transactions on Antennas and Propagation*, vol. 52, no. 1, pp. 59–65, 2004.

[15] R. Feick, H. Carrasco, M. Olmos and H. D. Hristov, 'PIFA input bandwidth enhancement by changing feed-plate silhouette,' *Electronics Letters*, vol. 40, no. 15, pp. 921–922, 2004.

[16] P. Salonen, M. Keskilammi and M. Kivikoski, 'Single-feed dual-band planar inverted-F antenna with U-shaped slot,' *IEEE Transactions on Antennas and Propagation*, vol. 48, no. 8, pp. 1262–1264, 2000.

[17] Z. N. Chen and M. Y. W. Chia, 'Broadband probe-fed notched plate antenna,' *Electronics Letters*, vol. 36, no. 7, pp. 599–600, 2000.

[18] Z. N. Chen, 'Impedance characteristics of a probe-fed L-shaped plate antenna,' *Radio Science*, vol. 36, no. 6, pp. 1377–1384, 2001.

[19] D. Nashaat, H. A. Elsadek and H. Ghali, 'Dual-band reduced-size PIFA antenna with U-slot for bluetooth and WLAN applications,' *IEEE International Symposium on Antennas and Propagation*, vol. 2, pp. 962–965, June 2003.

[20] K. Ogawa, T. Uwano and M. Takahshi, 'Shoulder-mounted planar antenna for mobile radio applications,' *IEEE Transactions on Vehicular Technology*, vol. 49, no. 3, pp. 1041–1044, 2000.

[21] C. R. Rowell and R. D. Murch, 'A capacitively loaded PIFA for compact mobile telephone handsets,' *IEEE Transactions on Antennas and Propagation*, vol. 45, no. 5, pp. 837–842, 1997.

[22] C. R. Rowell and R. D. Murch, 'A compact PIFA suitable for dual-frequency 900/1800-MHz operation,' *IEEE Transactions on Antennas and Propagation*, vol. 46, no. 4, pp. 596–598, 1998.

[23] T. Lo and Y. Hwang, 'Bandwidth enhancement of PIFA loaded with very high permittivity material using FDTD,' *IEEE International Symposium on Antennas and Propagation*, vol. 2, pp. 798–801, June 1998.

[24] Z. N. Chen, K. Hirasawa, K. W. Leung and K. M. Luk, 'A new inverted-F antenna with a ring dielectric resonator,' *IEEE Transactions on Vehicular Technology*, vol. 48, no. 4, pp. 1029–1032, 1999.

[25] G. A. Ellis and S. Liw, 'Active planar inverted-F antennas for wireless applications,' *IEEE Transactions on Antennas and Propagation*, vol. 51, no. 10, pp. 2899–2906, 2003.

[26] G. Lui and R. D.Murch, 'Compact dual-frequency PIFA designs using *LC* resonators,' *IEEE Transactions on Antennas and Propagation*, vol. 49, no. 7, pp. 1016–1019, 2001.

[27] K. L. Virga and Y. Rahmat-Samii, 'Low-profile enhanced-bandwidth PIFA antennas for wireless communications packaging,' *IEEE Transactions on Antennas and Propagation*, vol. 45, no. 10, pp. 1879–1888, 1997.

[28] P. Ciais, R. Staraj, G. Kossiavas and C. Luxey, 'Compact internal multiband antenna for mobile phone and WLAN standards,' *IEEE Microwave and Wireless Components Letters*, vol. 14, no. 4, pp. 148–150, 2004.

[29] R. Kronberger, H. Lindemeier, L. Reiter and J. Hopf, 'Multiband planar inverted-F car antenna for mobile phone and GPS,' *IEEE International Symposium on Antennas and Propagation*, vol. 4, pp. 2714–2717, June 1999.

[30] Z. N. Chen, 'Note on impedance characteristics of L-shaped wire monopole antenna,' *Microwave and Optical Technology Letters*, vol. 26, no. 1, pp. 22–23, 2000.

[31] G. Sanford and L. Klein, 'Recent developements in the design of conformal microstrip phased arrays,' *IEE Conference on Maritime and Aeronautical Satellites for Communication and Navigation*, pp. 105–180, March 1978.

[32] C. Wood, 'Improved bandwidth of microstrip antennas using parasitic elements,' *IEE Proceedings: Microwave, Antennas and Propagation*, vol. 127, no. 4, pp. 231–234, 1980.

[33] M. C. Huynh and W. Stutzman, 'Ground plane effects on planar inverted-F antenna (PIFA) performance,' *IEE Proceeedings: Microwave, Antennas and Propagation*, vol. 150, no. 4, pp. 209–213, 2003.

[34] A. S. Meier and W. P. Summers, 'Measured impedance of vertical antennas and effects of finite ground planes,' *Proceedings of the IEEE*, vol. 37, pp. 609–616, 1969.

[35] K. H. Awadalla and T. S. M. Maclean, 'Input impedance of a monopole antenna at the center of a finite ground plane,' *IEEE Transactions on Antennas and Propagation*, vol. 26, no. 2, pp. 244–248, 1978.

[36] M. M. Weiner, 'Monopole element at the center of a circular ground plane whose radius is small or comparable to a wavelength,' *IEEE Transactions on Antennas and Propagation*, vol. 35, no. 5, pp. 488–495, 1987.

[37] J. H. Richmond, 'Monopole antenna on circular disk,' *IEEE Transactions on Antennas and Propagation*, vol. 32, no. 12, pp. 1282–1287, 1984.

[38] J. Huang, 'The finite ground plane effect on the microstrip antenna radiation patterns,' *IEEE Transactions on Antennas and Propagation*, vol. 31, no. 4, pp. 649–655, 1983.

[39] A. K. Bhattacharyya, 'Effect of ground plane and dielectric truncation on the efficiency of a printed structure,' *IEEE Transactions on Antennas and Propagation*, vol. 39, no. 3, pp. 303–308, 1991.

[40] P. Moosavi and L. Shafai, 'Directivity of microstrip ring antennas and effects of finite ground plane on the radiation parameters,' *IEEE International Symposium on Antennas and Propagation*, vol. 2, pp. 672–675, June 1998.

[41] M. Geissler, D. Heberling and I. Wolff, 'Bandwidth and radiation properties of internal handset antennas,' *IEEE International Symposium on Antennas and Propagation*, vol. 4, pp. 2246–2249, 16–21 July 2000.

[42] A. T. Arkko and E. A. Lehtola, 'Simulation impedance bandwidths, gains, radiation patterns and SAR values of a helical and PIFA antenna on top of different ground planes,' *11th IEE International Conference on Antennas and Propagation*, vol. 2, pp. 732–735, April 2001.

[43] S. Saario, D. V. Thiel, J. W. Lu and S. G. O'Keefe, 'An assessment of cable radiation effects on mobile communications antenna measurements,' *IEEE International Symposium on Antennas and Propagation*, vol. 2, pp. 550–553, June 1997.

[44] C. Icheln, J. Ollikainen and P. Vainikainen, 'Reducing the influence of feed cables on small antenna measurements,' *Electronics Letters*, vol. 35, pp. 1212–1214, 1999.

[45] C. Icheln, M. Popov, P. Vainikainen and S. He, 'Optimal reduction of the influence of RF feed cables in small antenna measurements,' *Microwave and Optical Technology Letters*, vol. 25, no. 3, pp. 194–196, 2000.

[46] S. Pan, T. Becks, A. Bahrwas and I. Wolff, 'N antennas and their applications in portable handsets,' *IEEE Transactions on Antennas and Propagation*, vol. 42, no. 10, pp. 1475–1483, 1994.

[47] P. Salonen, L. Sydnheimo, M. Keskilammi and M. Kivikoski, 'Planar inverted-F antenna for wearable applications,' *IEEE 3rd International Symposium on Wearable Computers*, vol. 4, pp. 95–100, June 1999.

[48] L. Ukkonen, D. Engels, L. Sydänheimo and M. Kivikoski, 'Planar wire-type inverted-F RFID tag antenna mountable on metallic objects,' *IEEE International Symposium on Antennas and Propagation*, vol. 1, pp. 101–104, June 2004.

[49] T. W. Chiou and K. L. Wong, 'Design of compact microstrip antennas with a slotted ground plane,' *IEEE International Symposium on Antennas and Propagation*, vol. 2, pp. 651–654, 8–13 July 2001.

[50] R. Hossa, A. Byndas and M. E. Bialkowski, 'Improvement of compact terminal antenna performance by incorporating open-end slots in ground plane,' *IEEE Microwave and Wireless Componenets Letters*, vol. 14, no. 6, pp. 283–285, 2004.

[51] S. Rogers, J. Marsh, W. McKinzie and J. Scott, 'An AMC-based 802.11a/b antenna for laptop computers,' *IEEE International Symposium on Antennas and Propagation*, vol. 1, pp. 10–13, June 2003.

[52] W. E. Mckinzie and R. R. Fahr, 'A low-profile polarization diversity antenna built on an artificial magnetic conductor,' *IEEE International Symposium on Antennas and Propagation*, vol. 1, pp. 762–765, June 2002.

[53] R. Diaz, V. Sanches, E. Caswell and A. Miller, 'Magnetic loading of artificial conductors for bandwidth enhancement,' *IEEE International Symposium on Antennas and Propagation*, vol. 2, pp. 431–434, June 2003.

[54] Z. Du, K. Gong, J. S. Fu, B. Gao and Z. Feng, 'A compact planar inverted-F antenna with a PBG-type ground plane for mobile communications,' *IEEE Transactions on Vehicular Technology*, vol. 52, no. 3, pp. 483–489, 2000.

[55] P. Salonen, M. Keskilammi and M. Kivikoski, 'Dual-band and wide-band PIFA with U- and meanderline-shaped slots,' *IEEE International Symposium on Antennas and Propagation*, vol. 1, pp. 116–119, 8–13 July 2001.

[56] P. Teng, S. Fang and K. Wong, 'PIFA with a bent, meandered radiating arm for GSM/DCS dual-band operation,' *IEEE International Symposium on Antennas and Propagation*, vol. 1, pp. 107–110, 22–27 June 2003.

[57] A. F. Muscat and C. G. Parini, 'Novel compact handset antenna,' *11th IEE International Conference on Antennas and Propagation*, vol. 1, pp. 336–339, 17–20 April 2001.

[58] N. C. Karmakar, P. Hendro and L. S. Firmansyah, 'Shorting strap tunable single-feed dual-band PIFA,' *IEEE Microwave and Wireless Componenets Letters*, vol. 13, no. 1, pp. 13–15, 2003.

[59] A. K. Shackelford, S. Leong and K. F. Lee, 'Simulation of a probe-fed notched patch antenna with a shorting post,' *IEEE International Symposium on Antennas and Propagation*, vol. 2, pp. 708–711, 8–13 July 2001.

[60] Z. N. Chen and M. Y. W. Chia, 'A circular planar inverted-L antenna with vertical ground plane,' *Microwave and Optical Technology Letters*, vol. 35, no. 4, pp. 315–317, 2002.

[61] Z. N. Chen, K. Hirasawa and K. Wu, 'A broad-band sleeve monopole integrated into parallel-plate waveguide,' *IEEE Transactions on Microwave Theory and Techniques*, vol. 48, no. 7, pp. 1160–1163, 2000.

[62] C. Icheln and P. Vainikainen, 'Dual-frequency balun to decrease influence of RF feed cables in small antenna measurements,' *Electronics Letters*, vol. 36, pp. 1760–1761, 2000.

[63] W. A. T. Kotterman, G. F. Pedersen, K. Olesen and P. Eggers, 'Cable-less measurement set-up for wireless handheld terminals,' *Proceedings of Personal, Indoor, and Mobile Radio Communication*, vol. B, pp. 112–116, 2001.

[64] K. Fujimoto and J. R. James, *Mobile Antenna Systems*. Boston, MA: Artech House, 1994.

[65] J. Haley, T. Moore and J. T. Bernhard, 'Experimental investigation of antenna-handset-feed interaction during wireless product testing,' *Microwave and Optical Technology Letters*, vol. 34, no. 3, pp. 169–172, 2002.

[66] Z. N. Chen, N. Yang, Y. X. Guo and M. Y. W. Chia, 'An investigation into measurement of handset antennas,' *IEEE Transactions on Instrumentation and Measurement*, vol. 54, no. 3, pp. 1100–1110, June 2005.

[67] M. A. Jensen and Y. Rahmat-Samii, 'Performance analysis of antennas for hand-held transcievers using FDTD,' *IEEE Transactions on Antennas and Propagation*, vol. 42, no. 8, pp. 1106–1113, 1994.

[68] T. Hosoe and K. Ito, 'Dual-band planar inverted-F antenna for laptop computers,' *IEEE International Symposium on Antennas and Propagation*, vol. 3, pp. 87–90, June 2003.

[69] D. Liu and B. Gaucher, 'A branched inverted-F antenna for dual band WLAN applications,' *IEEE International Symposium on Antennas and Propagation*, vol. 4, pp. 2623–2626, June 2004.

[70] S. Rogers, J. Scott, J. Marsh and D. Lin, 'An embedded quad-band WLAN antenna for laptop computers and equivalent circuit model,' *IEEE International Symposium on Antennas and Propagation*, vol. 4, pp. 2588–2591, June 2004.

[71] K. Fukuchi, T. Ogawa, M. Ikegaya1, H. Tate and K. Takei, 'Small and thin structure plate type wideband antenna (3 GHz to 6 GHz) for wireless communications,' *IEEE International Symposium on Antennas and Propagation*, vol. 4, pp. 2615–2618, June 2004.

5

Planar Monopole Antennas and Ultra-wideband Applications

5.1 INTRODUCTION

Besides planar transmission-line antennas such as microstrip patch antennas, SPAs, and PIFAs or PILAs, dipoles and monopoles are the most basic types of antenna and have been used widely since Guglielmo Marconi, an Italian inventor, sent and received his first radio signal in Italy in 1895. A straight wire monopole vertically installed above a ground plane features simple structures but pure vertical polarization and horizontal omnidirectional radiation. The impedance bandwidth of simple thin-wire monopoles can be increased by modifying their geometry, such as thickening or loading or folding their wire elements. Typical designs include conical or skeletal conical, cage, and various loading monopoles.[1–4] However, as compared with the thin-wire monopoles, the conical or rotationally symmetric monopoles are much more bulky. Alternatively, planar elements have been used to replace the wire elements of the monopoles to broaden the impedance bandwidth.[5–11]

With planar radiators, the impedance bandwidth of monopole antennas can be broadened, typically up to more than 70 %, or even high-pass. Therefore, broadband planar antennas are becoming very attractive to promising developments such as software-defined radio systems, reconfigurable wireless communication systems and ultra-wideband (UWB) systems. Usually, the broadband design of antennas is much simpler than multiband design of narrowband antennas, such as planar fractal monopoles.[12] Figure 5.1 shows the geometry of a conventional triangle monopole and a fractal Sierpinski monopole. The former is a typical broadband antenna, called a 'bow-tie antenna' when it is used as a dipole. It is a variation of a finite-size conical antenna design. The Sierpinski monopole features multiband operation due to its log periodic resonant property. However, it may not be easy to control the performance, accurately and arbitrarily such as the center frequency and impedance bandwidth for each band, because the electromagnetic property of the fractal monopole is essentially determined by the uniformity or accuracy of the fractal structure.

Broadband Planar Antennas: Design and Applications   Zhi Ning Chen and Michael Y. W. Chia
© 2006 John Wiley & Sons, Ltd

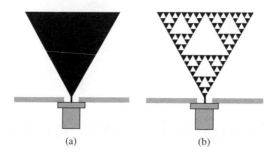

(a)   (b)

Figure 5.1 (a) A conventional planar triangle monopole, and (b) a Sierpinski monopole.

Increasing number of wireless systems are being integrated into a variety of portable devices, such as PDAs, cellphones and laptop computers. Typically, a laptop computer needs a WLAN system operating in the 2.45-GHz and 5–6-GHz bands. Due to the limited space available, embedded antennas capable of multiband/broadband operation are strongly desirable. Planar antennas are becoming the most important candidates for such applications.

UWB is an emerging wireless technology for commercial high-data-rate, short-range communications, radar systems, and measurement. The technology can be used within an ultra-wide spectrum but with an extremely low emission power level. For example, in 2002 the Federal Communications Commission (FCC) regulated the emission limits of −41.3 dBm/MHz for an allocated spectrum ranging from 3.1 GHz to 10.6 GHz.

In future, universal antenna solutions completely embedded into portable devices are desirable, which may cover frequencies from 800 MHz to 11 GHz or above in order to include all the existing wireless communication systems such as AMPC800, GSM900, GSM1800, PCS1900, WCDMA/UMTS (3G), 2.45/5.2/5.8-GHz-ISM, U-NII, DECT, WLANs, European HiperLAN I, II, UWB systems as well as narrowband wireless local loop.

The remainder of this chapter is divided into three main sections. Section 5.2 overviews the basic issues related to planar monopole antennas. A tutorial on the evolution of planar monopoles and their variations is first provided. Then, the method to calculate the lower edge frequencies of bandwidths of planar monopoles is given. After that, a roll monopole antenna is elaborated and discussed as a typical design of an antenna having a broad impedance and radiation performance.

Section 5.3 focusses on applications of planar antennas in ultra-wideband radio systems. First, the features of UWB radio systems are briefly introduced. The unique system requirements such as the emission limit masks are addressed. Then, the assessment criteria for antennas are discussed from a systems point of view. The design considerations for the antennas and source pulses are highlighted. Finally, some applications of planar antennas in UWB radio systems are exemplified.

This chapter ends with two case studies in Section 5.4. One is a directional antenna design for UWB radar systems, and the other is an omnidirectional antenna design for wireless communication applications in laptop computers.

5.2 PLANAR MONOPOLE ANTENNA

5.2.1 PLANAR BI-CONICAL STRUCTURE

Being simple but powerful antennas, monopoles with broad impedance and radiation bandwidths are desirable for many wireless devices. One way to enhance the impedance bandwidth of a simple thin-wire monopole is to increase its volume, such as the thickness of the radiating stem. However, in specific applications such as portable devices, bulky and heavy antennas are unacceptable. A tradeoff between broad bandwidth and compact volume is often made therefore the planar structure is used widely.

Figure 5.2 shows the evolution of a planar finite-size triangle monopole from the *infinite* bi-conical structure in (a). The latter is in principle frequency-independent when the inherent input impedance of the cone is the same as the characteristic impedance of the transmission line applied to feed the bi-conical structure. The angle of the cone determines the inherent input impedance of the cone. In (b), the bandwidth of the finite structure is limited because the cone with finite length cannot radiate efficiently at lower frequencies, so the lower edge of the bandwidth is determined by the largest dimension of the cone. These bi-conical structures can be solid or hollow.

To reduce the weight and possible wind-resistance of the bi-conical design in (b), a thin-wire structure was used as shown in (c). Alternatively, the planar structure shown in (d) can be used to reduce the space of the antenna installation. Comparing with the any 3-dimensional designs as shown in (a)–(c), the 2-dimensional triangle monopole as shown in (d) can readily be etched on a printed-circuit board (PCB) and integrated into other RF circuits on the PCB, where the planar monopole may be co-planar with the ground plane.

5.2.2 PLANAR MONOPOLES

Planar designs with various radiator shapes have been widely discussed and used.[5–11,13–23] Some of them are illustrated in Figure 5.3. Most planar monopoles feature very broad impedance bandwidths, typically of more than 70 % for VSWR = 2. For example, circular

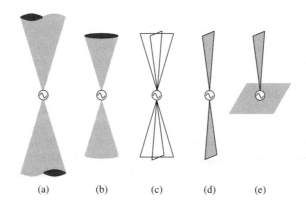

(a)   (b)   (c)   (d)   (e)

Figure 5.2   The evolution of planar finite triangle monopole from bi-conical structure.

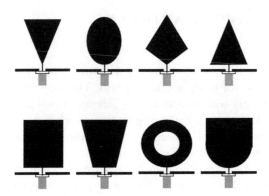

Figure 5.3 The variety of shapes of planar monopoles.

planar monopoles have high-pass impedance characteristics.[7,8] The feeding probe, the bottom of the planar radiator and the ground plane form the impedance transition. If good impedance matching occurs at multiple adjacent resonances simultaneously, the bandwidth will be widened.

Moreover, in contrast to the design of straight-wire monopoles, some unique design considerations should be taken into account for planar monopoles, such as the evaluation of the lower edge frequency of bandwidth (LEFBW) and the omnidirectivity of radiation. On one hand, the LEFBW for a broadband planar monopole may not be evaluated simply when the dominant resonant frequency is far from the center frequency of an operating bandwidth, due to the multiple resonances. On the other hand, with increasing operating frequency, the lateral dimension of the planar monopole will be electrically large so that the radiation from two vertical edges will be directional in horizontal planes.

The LEFBW

The operating frequency is one of the most important specifications in antenna design. For a single-resonance antenna, it can be evaluated by the dominant resonant frequency. For example, the operating frequency of a straight thin-wire monopole can be evaluated from $c/(4L_0)$, where c is the velocity of light and L_0 is the length of the thin-wire monopole. Therefore, the operating frequency and the achieved bandwidth are enough to describe the frequency response of the antenna under design, instead of the LEFBW. However, for a multi-resonance broadband antenna, the LEFBW and the achieved bandwidth will be useful for the monopole design instead of the operating frequency. Considering the increased size of the monopole, the dominant frequency is not determined simply by $c/(4L_0)$. For a thick monopole with circular cross-section, the formula is used to approximately evaluate the dominant resonant frequency.[24] For broadband planar monopoles, the evaluation becomes complicated because of their varying radiator shapes. Figure 5.4 shows the two types of planar trapezoidal monopoles.[17] Instead of the formula $4(H+h)$ for conventional thin-wire straight monopoles,

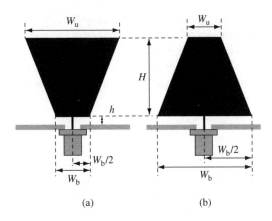

Figure 5.4 Planar trapezoidal monopoles.

equations 5.1 and 5.2 are both used to evaluate the LEFBW for trapezoidal planar monopoles:

$$F(\text{GHz}) = \frac{0.25 \times 300}{H + h + \frac{W_u + W_b}{4\pi}} \tag{5.1}$$

$$F(\text{GHz}) = \frac{0.25 \times 300}{\left\{ H^2 + \left[\frac{2\max(W_u, W_b) - W_u - W_b}{2} \right]^2 \right\} + h + \frac{W_u}{2\pi}} \tag{5.2}$$

where $\max(W_u, W_b) = W_u$ when $W_u \geq W_b$, or $\max(W_u, W_b) = W_b$ when $W_b \geq W_u$. Measurements demonstrated an 80% impedance bandwidth with an LEFBW of 1.36 GHz as against the calculated LEFBW of 1.61 GHz from equation 5.1, and 1.56 GHz from equation 5.2, where $H = 40$ mm, $h = 1$ mm, $W_u = 40$ mm, and $W_b = 30$ mm. The study also showed that for $W_u > W_b$, the calculated LEFBW from equation 5.2 can be much more accurate than that from equation 5.1 because the latter more accurately considers the effective radiating length of the planar sheet. The LEFBW is mainly determined by the effective radiating length including the side edges and top end of the radiating sheet.

As another example, a planar bow-tie monopole has been investigated and an equation presented for the evaluation of the LEFBW.[10] The modified equation has higher accuracy than the simple one in equation 5.1 based on the calculation of equivalent cylindrical monopoles.

5.2.3 ROLL MONOPOLES

As mentioned earlier, an important design consideration for planar monopoles is the change in the omnidirectivity of the radiation within the broad bandwidth. The non-axisymmetric structures of monopoles affect the omnidirectional radiation characteristics in horizontal planes, especially at the higher operating frequencies within the bandwidth. This degraded radiation performance more or less offsets the advantage of the reduction in antenna volume. In vertical planes, the directions of maximum beam vary and the gain decreases when the operating frequency increases. For a cylindrical monopole, a sleeve around the stem can be used to improve the radiation performance over a broad bandwidth.[25,26] This technique has not been used in planar designs.

The Roll Monopole

To enhance the radiation performance of a broadband planar monopole and keep it compact, the concept of a roll monopole has been presented.[27] Figure 5.5 shows the geometry. The monopole is made by uniformly rolling a perfectly electrically conducting (PEC) sheet with size $W \times H = 75\,\text{mm} \times 50\,\text{mm}$. The trace of its cross-section is described by

$$r = r_o + \alpha\theta \tag{5.3}$$

where r_o is the inner radius, α is the constant related to the spacing between two adjacent rolled layers, and the angle θ ranges from $0°$ to $360° \times N$. The term N is the number of roll turns and it might not be an integer.

As an example, a roll monopole with $r_0 = 4\,\text{mm}$, $\alpha = 0.5/360°$, and $N = 2.5$ was tested. The spacing between two adjacent layers is about $0.5\,\text{mm}$. This monopole is vertically mounted at the center of a $320\,\text{mm} \times 320\,\text{mm}$ ground plane. The bottom of the monopole is parallel to the ground plane with a feed gap $g = 1\,\text{mm}$. A 1.2-mm thick probe, an extension of the inner conductor of the 50-Ω coaxial line, fed the bottom of the roll at the point $(r_0, \theta = 0°)$ through the ground plane. The location of the feed point can be optimized for broad bandwidth.

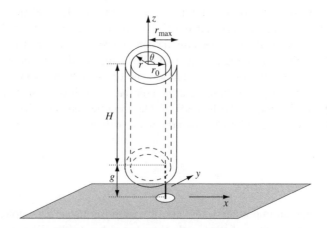

Figure 5.5   Geometry of a roll monopole. (Reproduced by permission of IEEE.[27])

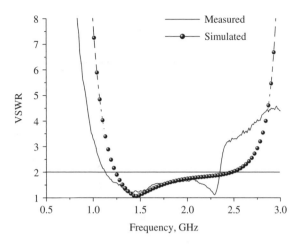

Figure 5.6 Comparison of simulated and measured VSWR. (Reproduced by permission of IEEE.[27])

Figure 5.6 plots the simulated and measured VSWR. An electromagnetic simulator (Zeland IE3D) was used to simulate the monopoles numerically, based on the method of moments. A broad impedance bandwidth of more than 70 % for VSWR = 2 was obtained. The simulated and measured LEFBWs are 1.21 GHz and 1.12 GHz, respectively. The structure with a planar helical cross-section introduces additional inductance, and the electromagnetic coupling between the rolled layers adds additional capacitance. Such a reactive network produces good impedance matching over a broad bandwidth.

The radiation patterns and gain for the roll monopole were simulated and measured in $\phi = 0°$, $\phi = 90°$ and $\theta = 90°$ planes. The radiation patterns are consistent within the bandwidth and quite similar to conventional cylindrical monopoles. For instance, Figure 5.7 shows the radiation patterns at 1.4, 1.8 and 2.2 GHz in the x-z plane. Good agreement has been demonstrated between the simulated and measured radiation patterns for E_θ components (or co-polarized components) in the top half space. However, there are large differences between the simulated and measured results for E_ϕ components because in the simulations, an infinite ground plane is used as against the finite ground plane in the measurements. Thus, the radiation patterns are simulated only above the ground plane in the ϕ-cuts, and the gain is slightly higher than the measured ones.

Figure 5.8 shows that, due to the almost symmetrical structure of the roll monopole, the measured radiation patterns in the x-y plane for the E_θ components are quite omnidirectional across the impedance bandwidth. Therefore, the radiation performance of a roll monopole is superior to that of a planar monopole from this point of view. Figure 5.9 compares the simulated and measured gain across the bandwidth, where the maximum gain was measured in $\phi = 0°$ and $\phi = 90°$ planes, and the simulated gain is the maxima in the two planes. Within the frequency band ranging from 1.2 GHz to 2.2 GHz, the roll monopole achieves quite stable and high gain of between 3.2 dBi and 4.6 dBi. The measured gain is about 0.5 dBi lower than simulated ones because of the use of the finite ground plane in the tests. The directions of maximum beam vary between 56° and 63° within the bandwidth of 1.2–2.2 GHz.

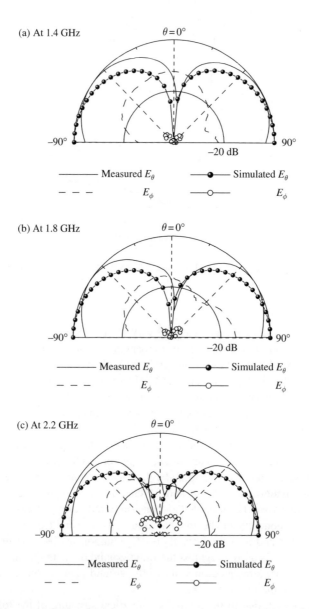

Figure 5.7 Simulated and measured radiation patterns across the bandwidth at x–z planes. (Reproduced by permission of IEEE.[27])

Comparison with Other Monopoles

In this section, characteristics of a thin-wire monopole, a thick cylindrical monopole, a planar monopole and a roll monopole are examined and compared in terms of impedance bandwidth, radiation pattern, gain and beam-maximum direction. Four typical monopoles are selected for comparison, as shown in Figure 5.10.

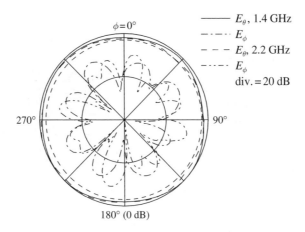

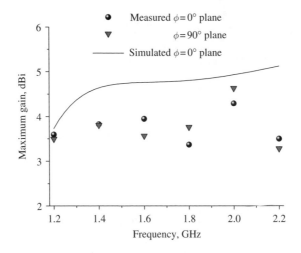

Figure 5.8 Simulated and measured radiation patterns across the bandwidth at x–y planes. (Reproduced by permission of IEEE.[27])

Figure 5.9 Simulated and measured maximum gain across the bandwidth. (Reproduced by permission of IEEE.[27])

   All the monopoles were vertically mounted at the center of a 320 mm×320-mm ground plane and an RF signal cable was connected to a 50-Ω SMA connector under the ground plane. To suppress the additional radiation from the RF feeding cable in tests, the RF cable was enclosed by an absorber layer and kept perpendicular to the ground plane. The bottom of the monopoles is parallel to the ground plane with a feed gap of 1 mm. The height of all the monopoles is 50 mm. The diameters of Monopoles 1 and 2 are 2 mm and 10 mm, respectively. Monopole 3 is a planar monopole with a rectangular 0.2-mm thick copper sheet of width 75 mm and the feed point located at $S = 30$ mm for maximum impedance bandwidth. Monopole 4 was formed by uniformly rolling Monopole 3. The trace of its

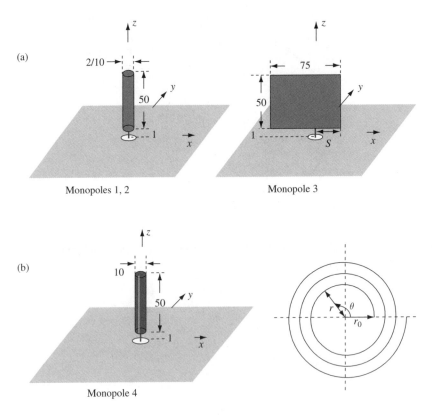

Figure 5.10 Geometry of monopoles under comparison (dimensions in millimeters). (Reproduced by permission of IEE.[28])

cross-section can be described by equation 5.3. In the measurements, the parameters of the roll are $r_0 = 4$ mm, $\alpha = 0.5/360°$ and $N = 2.5$. The distance between the two adjacent layers is 0.5 mm.

The impedance and radiation characteristics of these monopoles were compared experimentally and numerically.[28] The input impedance was measured by using an HP8510C Network Analyzer and the radiation patterns were tested in an anechoic chamber using an orbit far-field measurement system. The simulations were implemented using an electromagnetic simulator (IE3D) software package based on the method of moments.

Figures 5.11 details the comparison of the simulated and measured VSWR against frequency. The measured 2:1 VSWR impedance bandwidths show that, as against the 25 %, 40 % and 53 % bandwidths of Monopoles 1–3, the roll monopole (Monopole 4) has the largest bandwidth of 71 %. Furthermore, the frequencies of the lower edge of the bandwidths vary from 1.12 GHz to 1.38 GHz. The coupling between the layers of the roll leads to parasitic capacitance and the spiral structure introduces parasitic inductance. The parasitic capacitance and inductance result in additional resonances so that the compact Monopole 4 is able to achieve a broader impedance bandwidth than Monopole 3.

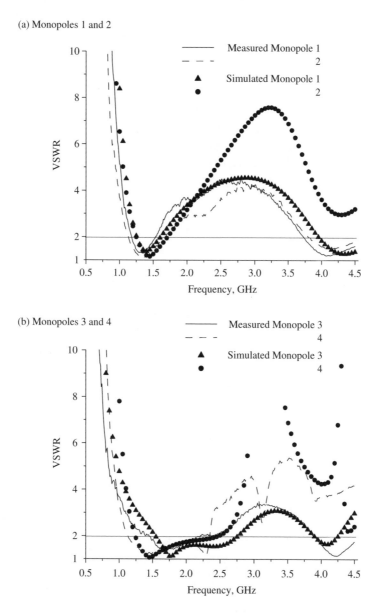

Figure 5.11 Comparison of simulated and measured VSWR for monopoles. (Reproduced by permission of IEE.[28])

The radiation properties for Monopoles 1–4 were examined in x–z, y–z and x–y planes. The results for Monopoles 1 and 2 show that the cylindrical monopoles with radii of 2 mm and 10 mm have almost the same radiation performance. However, the planar monopole (Monopole 3) with asymmetrical geometry, which is usually used to increase the bandwidth of monopoles with a reduced size, suffers from degradation of the radiation patterns, as

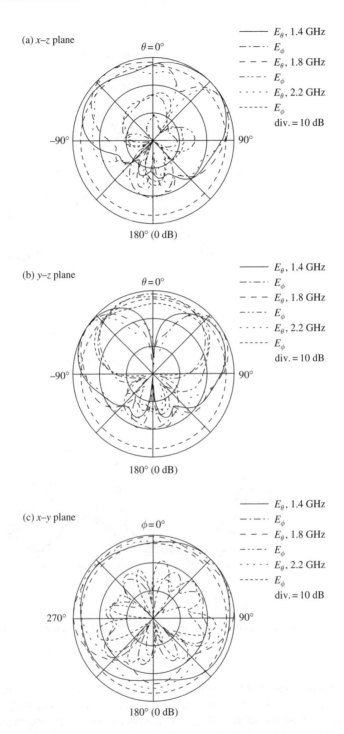

Figure 5.12   Comparison of radiation patterns for Monopole 3. (Reproduced by permission of IEE.[28])

shown in Figure 5.12. The radiation patterns were measured at 1.4, 1.8 and 2.2 GHz. In the
$x–z$ plane, the E_ϕ component levels are still low but the radiation of the E_θ components
features high levels and end-fire patterns. The asymmetry of the patterns ($x–z$ plane) for the
E_θ components is caused mainly by the asymmetrical feed configuration. In the $y–z$ plane,
as against the end-fire patterns for the E_ϕ components, the patterns for the E_θ components
are typical broadside. However, the E_ϕ component levels are almost the same as the E_θ
components because at the radiating sheet, both the x and z-directed electric currents are
excited simultaneously. Also, the severe asymmetry of the monopole causes degradation of
the omnidirectional radiation patterns of the E_θ components in the horizontal ($x–y$) plane,
as shown in Figure 5.12(c). This degradation becomes worse when the operating frequency
increases.

The radiation patterns for Monopole 4 were also measured at 1.4, 1.8 and 2.2 GHz.
Figures 5.13(a) and (b) demonstrate the same radiation properties as the cylindrical

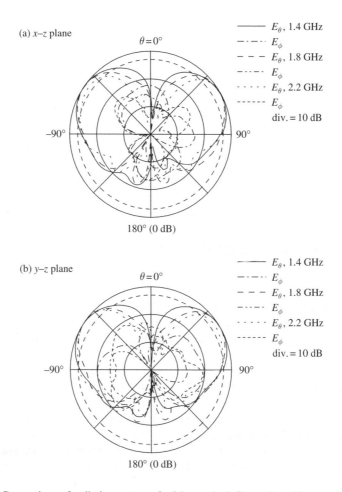

Figure 5.13 Comparison of radiation patterns for Monopole 4. (Reproduced by permission of IEE.[28])

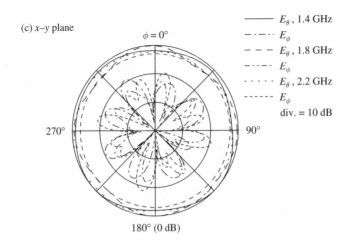

Figure 5.13   (Continued).

monopoles. Figure 5.13(c) shows that, due to the symmetrical-like structure, the radiation patterns for both E_θ and E_ϕ components are almost omnidirectional when considering the impact of the ground plane on the radiation performance observed in the measurements on Monopoles 1, 2 or 3.

From the detailed comparisons of the radiation properties of the four monopoles, it can be concluded that the roll monopole antenna with a compact structure has a broad impedance bandwidth as a planar monopole and stable radiation properties like a cylindrical monopole.

5.2.4 EMC FEEDING METHODS

Planar radiators can be fed directly by probes as discussed above. By adjusting the feed gap and location of the feed point, good impedance matching can be achieved. Furthermore, planar monopoles may be excited electromagnetically, namely by electro-magnetic coupling (EMC) between the planar radiator and the feeding strips fed by the probes.[18–20,22] This feeding method has additional design freedoms for broadband impedance matching.

Figure 5.14 shows three designs with EMC feeding structures. In (a), a square radiator is electromagnetically coupled to a rectangular strip fed by a short probe at its bottom.[19] By optimizing the size and location of the strip, the 2:1 VSWR impedance bandwidth reached up to 77 %. Figures 5.14(b) and (c) show triangular and annular planar monopoles with EMC feeding structures, which have the enhanced impedance bandwidths of >2.7:1 and 4.7:1 for 3:1 VSWR.[20,21] Compared with the planar monopole having a simple feeding structure, EMC planar monopoles can achieve broader bandwidths by virtue of the additional design freedom.

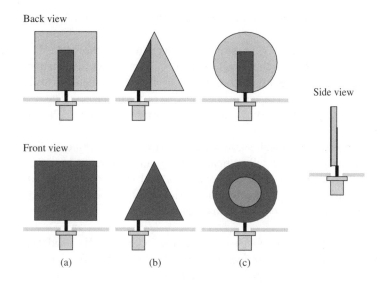

Figure 5.14 Planar monopoles with EMC feeding structures.

5.3 PLANAR ANTENNAS FOR UWB APPLICATIONS

5.3.1 ULTRA-WIDEBAND TECHNOLOGY

Since the 1970s, UWB radio technology has been investigated widely and developed for wireless applications.[29–33] Recently, much effort has been devoted to commercial short-range UWB systems. The commercialization of UWB radio devices followed swiffly on the recent endorsement of the spectra by the FCC,[34] and later international efforts towards a widely compatible regulatory framework. Potential commercial opportunities for UWB radio technology are expected for communications, imaging, ranging and localization. UWB devices are typically used to transmit and/or receive signals with very short pulses, which may be modulated in UWB communication systems. The extremely short pulses in the time domain usually occupy ultra-wide bandwidths in the frequency domain. The commercial UWB devices definited by the FCC include imaging systems, vehicular radar systems, communications and measurement systems.

Imaging Systems provide for the operation of ground-penetrating radar (GPR) systems and other imaging devices subject to certain frequency and power limitations. Imaging systems include:

- *Ground-penetrating radar systems.* GPR systems must be operated below 960 MHz or in the frequency band of 3.1–10.6 GHz. These systems operate only when in contact with, or in close proximity to, the ground for the purpose of detecting or obtaining images of buried objects. The energy from the GPR system is intentionally directed down into the ground for this purpose.
- *Wall imaging systems.* These must be operated below 960 MHz or in the frequency band of 3.1–10.6 GHz. Wall-imaging systems are designed to detect the location of objects

contained within a wall, such as a concrete structure, the side of a bridge or the wall of a mine.

- *Through-wall imaging systems.* These must be operated below 960 MHz or in the frequency band of 1.99–10.6 GHz. Through-wall imaging systems detect the location or movement of persons or objects located on the other side of a structure such as a wall.
- *Medical imaging systems.* These must be operated in the frequency band of 3.1–10.6 GHz. A medical imaging system may be used for a variety of health applications in humans and animals.
- *Surveillance systems.* Although surveillance devices are not technically imaging systems, for regulatory purposes they will be treated in the same way as through-wall imaging and will be permitted to operate in the frequency band of 1.99–10.6 GHz. Surveillance systems operate as security fences by establishing a stationary RF perimeter field and detecting the intrusion of persons or objects in that field.

Vehicular Radar Systems provides for operation in the 24 GHz band using directional antennas on terrestrial transportation vehicles, provided the center frequency of the emission and the frequency at which the highest radiated emission occurs are greater than 24.075 GHz. These devices are able to detect the location and movement of objects near a vehicle, enabling features such as near-collision avoidance, improved airbag activation, and suspension systems that respond better to road conditions.

Communications and Measurement Systems provide for the use of a wide variety of other UWB devices, such as high-speed home and business networking devices as well as storage tank measurement devices under Part 15 of the Commission's rules subject to certain frequency and power limitations. The devices must operate in the frequency band of 3.1–10.6 GHz. The equipment must be designed to ensure that operation can occur only indoors, or it must consist of hand-held devices that may be employed for such activities as peer- to-peer operation.

With their ultra-wide bandwidth or short duration of pulses, UWB systems offer potential opportunities for high-resolution radar imaging, rejection of the multipath cancellation effect, transmission of high-data-rate signals, and coding for security especially in multi-user network applications.[35-38] In addition, UWB radio systems may transmit and receive short time pulses without any carrier, or modulated short pulses with carrier.

Carrier-free UWB radio systems usually employ short pulses typically of the order of sub-nanoseconds such that one or a few pulses may occupy an extremely broad bandwidth, typically larger than 20 % or 500 MHz. Such systems are capable of providing low system complexity and low costs, because of the direct transmission and reception of the pulsed signals and fewer RF devices in their front-ends compared with conventional narrowband radio systems.

On the other hand, UWB radio/radar systems may possibly interfere with existing electronic systems since the short pulses or modulated pulses occupy extremely wide spectra which may overlap the bands utilized by other existing electronic systems. Owing to potential interference, the FCC regulated the emission limits, for instance, the effective isotropic radiated power (EIRP) levels of −41.3 dBm/MHz for the allocated 7.5-GHz bandwidth (3.1–10.6 GHz, termed the UWB band) for the unlicenced use of commercial UWB communication systems. These emission limits will be the crucial consideration in the design of source pulses and antennas for UWB radio systems. The radiated power spectral density (PSD) shaping can be controlled by selecting source pulses, and tailored by designing transmitting antennas.

In addition, the emission limits indicate that UWB systems may operate across a 7.5-GHz or a 110 % 10-dB fractional bandwidth. As a result, the transmission and reception of pulsed signals in UWB systems are distinct from conventional narrowband systems.

- First, the design of the source pulses may significantly affect the performance of pulsed UWB systems. By properly selecting the source pulse, the radiated power within the UWB band can be maximized and still comply with the required emission limits without the need for filters before transmitting antennas.
- Second, the waveform of the pulses arriving at a receiver usually do not resemble the waveform of the source pulses at a transmitter in a pulsed UWB system. Transmit/receive antennas with frequency-dependent transfer responses act as temporal differentiators/integrators or spectral/spatial filters if the bandwidth of the antenna in terms of magnitude and phase cannot cover the operating bandwidth well. The selection of a template for correlation detection at a receiver strongly depends on the characteristics of both source pulses and transmit/receive antennas.
- Third, antennas in UWB systems should be analysed and modeled in both time domain (TD) and frequency domain (FD).[39–43]
- Fourth, the antennas should be designed and assessed from an overall systems point of view, and not just as individual antenna elements.[39,44–49]

5.3.2 CONSIDERATIONS FOR UWB ANTENNAS AND SOURCE PULSES

In comparison with conventional narrowband/broadband systems, two essential and special design considerations for antennas in UWB systems should be emphasized. One is that the power spectral density (PSD) shaping of the radiated signals should conform to the emission limit masks to avoid possible interference with other electronic systems. The other point is that the source pulses and transmit/receive antennas should be evaluated collectively from a systems point of view.

In this section, the discussion focusses on antennas for UWB radio systems. First, a frequency-dependent transmission equation based on the Friis transmission formula is employed to describe transmit/receive antenna systems. Source pulses and transmitting antennas are studied and optimized in the context of the emission limits. Next, transmit/receive antenna systems are assessed in terms of pulse fidelity and system transmission efficiency. Thin-wire dipoles with narrow bandwidths and planar dipoles with broad bandwidths are exemplified in the discussion. The numerical methods used in the discussions include a finite difference time domain (FDTD) method, a time-domain integral equation (TDIE) method, and the method of moments (MoM).

Description of Antenna Systems

Consider a typical transmit/receive antenna setup in a UWB radio system, as shown in Figure 5.15. The Friis transmission formula can be used to relate the output power

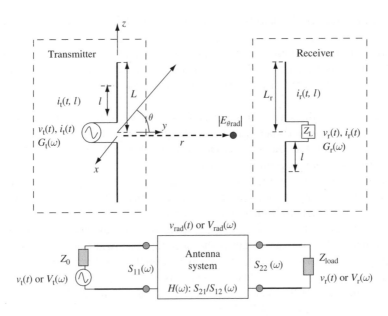

Figure 5.15   Transmit/receive antenna system. (Reproduced by permission of IEEE.[39])

of the receiving antenna to the input power of the transmitting antenna as shown in equation 5.4, where it is assumed that each antenna is in the far-field zone of the other:

$$\frac{P_r}{P_t} = \left(1 - |\Gamma_t|^2\right)\left(1 - |\Gamma_r|^2\right) G_t G_r |\hat{\rho}_t \cdot \hat{\rho}_r|^2 \left(\frac{\lambda}{4\pi r}\right)^2 \tag{5.4}$$

where P_t and P_r are, respectively, the time-average input power of the transmitting antenna and the time-average output power of the receiving antenna; Γ_t and Γ_r are, respectively, the return loss at the input of the transmitting antenna and the output of the receiving antenna; G_t and G_r are, respectively, the gain of the transmitting and receiving antennas; $\hat{\rho}_t$ and $\hat{\rho}_r$ are the polarization matching factors between the transmitting and receiving antennas; λ is the operating wavelength; and r is the distance between the transmitting and receiving antennas. The gain of the antennas, G_t and G_r, are functions of orientation (θ, ϕ). This formula is for cases where the parameters in equation 5.4 within the operating frequency range are unchanged. For general cases, equation 5.4 can be modified as equation 5.5 because the response between transmitting and receiving antennas is frequency-dependent:

$$\frac{P_r(\omega)}{P_t(\omega)} = \left(1 - |\Gamma_t(\omega)|^2\right)\left(1 - |\Gamma_r(\omega)|^2\right) G_t(\omega) G_r(\omega) |\hat{\rho}_t(\omega) \cdot \hat{\rho}_r(\omega)|^2 \left(\frac{\lambda}{4\pi r}\right)^2 . \tag{5.5}$$

If a transfer function $H(\omega)$ is defined to describe the relation between the source and output signal (voltage) $[V_t(\omega)/2]^2/2\ (=P_t(\omega)Z_0)$ and $[V_r(\omega)/2]^2/2\ (=P_r(\omega)Z_{load})$, equation 5.5 can be simplified as

$$H(\omega) = \frac{V_r(\omega)}{V_t(\omega)} = \left| \sqrt{\frac{P_r(\omega)}{P_t(\omega)} \frac{Z_{load}}{4Z_0}} \right| = |H(\omega)|e^{-j\phi(\omega)};$$

$$\phi(\omega) = \phi_t(\omega) + \phi_r(\omega) + \frac{\omega r}{c} \qquad (5.6)$$

where c denotes the velocity of light, and $\phi_t(\omega)$ and $\phi_r(\omega)$ are, respectively, the phase variations caused by the transmitting and receiving antennas. The transfer characteristics of the transmit/receive antenna system are determined by the performance parameters of the two antennas, such as impedance matching, gain, polarization matching, the distance between the antennas, the operating frequency, as well as the orientation of the antennas if the characteristics of an RF channel are completely ignored. Therefore, $H(\omega)$ can be used to describe the antenna system, which may be dispersive.

The antenna system can be considered also as a two-port network as shown in Figure 5.15, so the transfer function can be measured in terms of the parameter S_{21} when the source impedance and loading are matched to the transmitting and receiving antennas, respectively. Therefore, the measurable parameter S_{21} or $H(\omega)$ can be used to evaluate the performance of an antenna system.

Similarly, the relation between the radiated electric fields and source pulses at the transmitting antenna can be expressed as

$$E_{rad}(\omega) = H_{rad}(\omega)V_t(\omega) = \hat{a}|H_{rad}(\omega)|e^{-j\phi_{rad}(\omega)}V_t(\omega);$$

$$\phi_{rad}(\omega) = \phi_t(\omega) + \frac{\omega r}{c}. \qquad (5.7)$$

The transfer function $E_{rad}(\omega)$ is a vector with polarization direction $\hat{a}$ of the transmitting antenna and determined by the characteristics of the transmitting antenna, such as impedance matching, gain and orientation of the observation point. $V_t(\omega)$ is the spectrum of a source signal (voltage). Therefore, the radiation transfer function can be used to evaluate the radiated PSD for evaluation of the emission limits.

Emission Limit Masks

Emission level limits are measured in terms of effective isotropic radiated power (EIRP) levels within the available spectra and the measurement method specified by regulatory bodies such as FCC in the USA. Usually, the emission levels must be lower than noise, namely -41.3 dBm/MHz. For instance, the emission levels within 0.99–1.66 GHz must be lower than -75 dBm/MHz because the GPS service falls within this range. More information on emission limit masks and measurement methods can be obtained from the regulatory bodies; they vary with different countries and organizations. The low emission limits differentiate the design considerations for source pulses and transmitting antennas in UWB radio systems from conventional narrowband ones.

In practice, radio systems can utilize the UWB band in a variety of ways. For instance, multi-band (single-/multi-carrier) and single-band schemes can be used. To comply with the emission limit masks, the design considerations for source pulses and transmitting antennas are subject to the specific system schemes.

- *Multi-band scheme.* Under this scheme, the available UWB band can be divided into several sub-bands. Each of the source pulses is shaped to occupy only one sub-band. For example, Figure 5.16(a) shows the scheme of 15 uniform sub-bands for the 7.5-GHz band, where the 10-dB bandwidths are of 500 MHz. Figure 5.16(b) displays a Gaussian pulse $e^{-(t/a)^2}$ with $\sigma = 1366$ ps, which is modulated by sine-wave signals with frequencies of $(3.35 + n \times 0.5)$ GHz ($n = 0, 1, 2, \ldots, 14$). It is clear that possible interference with other systems outside the UWB band and between the sub-bands can be suppressed if the pulse parameter σ and the modulation frequencies are properly set. Another advantage of this scheme is to avoid the possible interference appearing within the UWB band because it allows UWB radio system users to suspend any sub-band separately, within which the interference with other systems becomes severe. For example, the transmission of a pulse covering the sub-band of 5.6–6.1 GHz will halt when UWB radio systems have a strong interference with WLAN users. Consequently, this scheme is good for interference considerations, but it greatly increases the complexity of system design.
- *Single-band scheme.* The alternative single-band scheme was initially proposed for UWB radio/radar technology. The source pulses, which usually have a very short duration, are shaped so that their spectra occupy as wide as possible a range within the UWB band for high data rates and good signal-to-noise ratio.

From equation 5.7, it can be seen that there are at least two methods to meet the emission limit masks by properly designing the source pulses and antennas. One method is to optimize the spectra $V_t(\omega)$ of source signals directly to make the 10-dB bandwidth of the source signal narrower than the UWB band when the antenna system has a constant unchanged radiation transfer function $H_{rad}(\omega)$ within the UWB band. This involves two scenarios. One is that the 10-dB bandwidth falls fully into the UWB band by properly selecting the source pulses. Otherwise, the 10-dB bandwidth spectrum of the pulse can be shifted into the UWB band by modulating the pulse with a proper sine-wave signal (carrier). Both cases will make antenna design relatively easy.

The other method is to tailor the spectra $V_t(\omega)$ of source signals by using the filtering function of $H_{rad}(\omega)$; that is, to control $H_{rad}(\omega)V_t(\omega)$ if $V_t(\omega)$ does not meet the emission limit mask. Using this method, the transmitting antenna acts not only as a radiator but also as a filter, which is designed to suppress the unwanted radiation outside the UWB band or in the specific band. This will make antenna design complicated.

Source Pulses

Assuming that the bandwidth of a transmitting antenna is broad enough so that, within the UWB band, $H_{rad}(\omega)$ is constant, then in principle all pulses with spectra (wider than 500 MHz stipulated by FCC) falling into the UWB band can be used as signals. However, in

(a) Frequency domain

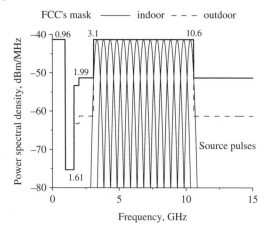

(b) Time domain

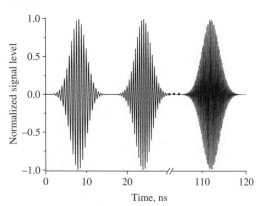

Figure 5.16 Pulses and spectra in a multi-band scheme. (Reproduced by permission of IEEE.[39])

practice, only pulses that can be easily generated and controlled, and with low power consumption (no direct-current component) are selected. Owing to their unique temporal and spectral properties, a family of Rayleigh (differentiated Gaussian) pulses, $v_n(t)$ or $\bar{v}_n(\omega)$, is widely used in UWB systems:

$$v_n(t) = \frac{\mathrm{d}^n}{\mathrm{d}t^n}\left[e^{-\left(\frac{t}{\sigma}\right)^2}\right]; \quad \bar{v}_n(\omega) = (j\omega)^n \sigma\sqrt{\pi} e^{\left(\frac{\omega\sigma}{2}\right)^2} \tag{5.8}$$

where the pulse parameter σ stands for the time when $v_0(\sigma) = 1/e$. The pulse duration T is defined as the interval between the start and the end of the pulse, where the values $|v_n(t = \pm T/2)|$ decreases from the normalized peak value to e^{-9} as shown in Figure 5.17(b). The figure displays the Gaussian pulse $v_0(t)$ and first-order Rayleigh pulse $v_1(t)$ for $\sigma = 20$,

(a) Frequency domain

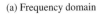

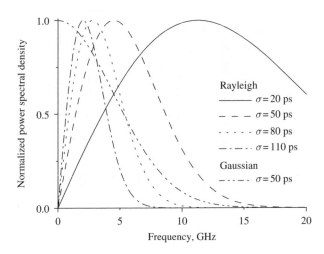

(b) Time domain

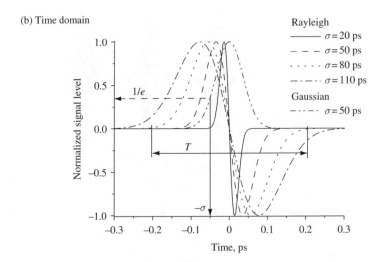

Figure 5.17 Pulses and spectra in a single-band scheme. (Reproduced by permission of IEEE.[39])

50, 80 and 110 ps. $v_1(t)$ is a monocycle pulse, which is easily generated by RF circuits and does not generate any DC component in the frequency domain.

Calculations show that the 10-dB bandwidths of the first-order Rayleigh pulses with $\sigma > 61$ ps ($T > 305$ ps) are <7.5 GHz, as shown in Figure 5.17(a). However, their spectra do not fall fully into the UWB band defined by the FCC. Some higher-order Rayleigh pulses can match the UWB band directly, such as the fourth-order Rayleigh pulses with 67 ps $< \sigma < 76$ ps, fifth-order Rayleigh pulses with 72 ps $< \sigma < 91$ ps, and sixth-order Rayleigh pulses with 76 ps $< \sigma < 106$ ps.

Transmitting Antenna and PSD

To examine the effects of source pulses and transmitting antennas on radiated PSD shaping, two types of antennas with narrow and broad impedance bandwidths are exemplified. Center-fed thin-wire dipole antennas with narrow bandwidths are considered first. A thin-wire straight dipole with lengths $L = 11$ mm and having a 0.3-mm radius was numerically simulated using the Time domain integral equation (TDIE) method. The resonant frequency is 5.85 GHz with a 10-dB bandwidth of 25 % (for $|S_{11}|$) as shown in Figure 5.18. The normalized radiated transfer function $|H_{rad}(\omega)|$ is also illustrated. Three typical first-order Rayleigh pulses with $\sigma = 30$, 45 and 80 ps are used as source pulses. The radiated fields are the co-polarized components $|E_{rad}|$ in the direction of $\theta = 90°$ and at a distance $r = 1960$ mm, as shown in Figure 5.15. The comparison between the spectra of source and radiated pulses normalized to -41.3 dBm/MHz is depicted in Figure 5.18.

From the transfer function $|H_{rad}(\omega)|$ shown, it is readily observed that the dipole acts as a high-pass filter within the UWB band. Thus, the tailored radiated spectrum of the short pulse cannot fully meet the emission limit mask when the pulse has high emission levels at frequencies higher than 10.6 GHz, as shown in Figure 5.18(a). In contrast, (c) reveals that the radiated spectrum of the longest pulse also does not meet the emission limit masks because of its high emission levels at frequencies lower than 3.1 GHz. Figure 5.18(b) evidently demonstrates that the radiated spectrum can completely comply with the specific emission limit mask (indoor) by properly selecting the source pulses for a given transmitting antenna.

Another important parameter, the efficiency η of the transmitting antenna, can be evaluated from

$$\eta = \frac{\int_0^\infty P_t(\omega)\left(1 - |S_{11}(\omega)|^2\right)d\omega}{\int_0^\infty P_t(\omega)d\omega} \times 100\%. \tag{5.9}$$

Both source pulse and transmitting antenna determine the efficiency. The calculated efficiency is about 53 % for the case shown in Figure 5.18(b), where the spectrum conforms to the emission limit mask well but the antenna is a narrowband design with high return loss in most of the UWB band.

Figure 5.19 compares the waveforms of the radiated electric fields with $\sigma = 30$, 45 and 80 ps, or $\sigma_{ant}/\sigma = 1.22$, 0.82 and 0.46. The parameter $\sigma_{ant} = L/c$ indicates the time for light to travel the length L of the antenna arm at a velocity of 3×10^8 m/s. The figure shows the waveforms of the radiated pulses distorted by the highpass filtering of the antennas in the frequency domain.[50] Distortion in the time domain can be attributed mainly to the reflection appearing at the ends (including the input) of the dipole, as shown in Figure 5.20. The pulses radiated from the ends of the dipole (namely points O, B and C) arrive at the receiving antenna located at point A (in the far-field zone) through paths of different lengths. The length differences between the paths also cause time delay in the time domain or phase difference in the frequency domain.

As an example, the effect of impedance matching between the source and input of the transmitting antenna on the waveforms of the radiated pulses is explained in the frequency domain. The relation between the input current $I(\omega)$, voltage $U(\omega)$ and impedance $Z(\omega)$ of a thin-wire dipole is

$$I(\omega) = \frac{U(\omega)}{Z(\omega)}. \tag{5.10}$$

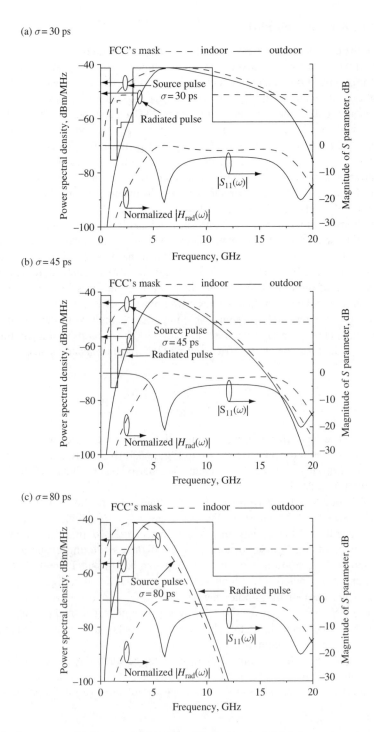

Figure 5.18 Power spectral density shaping of radiated electrical fields by a narrowband thin-wire dipole driven by first-order Rayleigh source pulses. (Reproduced by permission of IEEE.[39])

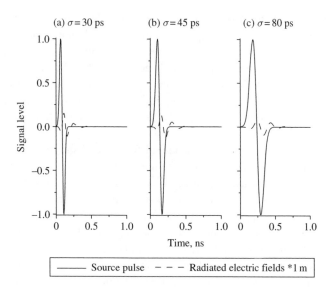

Figure 5.19 Comparison of waveforms of source pulses and the radiated electric fields ($\times 1$ m): (a) $\sigma = 30$ ps; (b) $\sigma = 45$ ps; (c) $\sigma = 80$ ps. (Reproduced by permission of IEEE.[39])

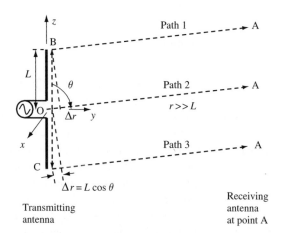

Figure 5.20 Multipath model of the radiation from a transmitting antenna. (Reproduced by permission of IEEE.[39])

The corresponding inverse Fourier transform is

$$i(t) = F^{-1}\left[\frac{U(\omega)}{Z(\omega)}\right] = \begin{cases} F^{-1}[j\omega CU(\omega)] \cong C\frac{dU(t)}{dt}, & \text{for } Z(\omega) \cong -\frac{j}{\omega C} \\ F^{-1}\left[\frac{U(\omega)}{R}\right] = \frac{U(t)}{R}, & \text{for } Z(\omega) \\ F^{-1}\left[\frac{U(\omega)}{Z(\omega)}\right], & \text{otherwise} \end{cases} \qquad (5.11)$$

It is clear that the waveform of the input current is related to the frequency-dependent input impedance of the dipole. For $\sigma_{ant}/\sigma << 1$ ($L << \lambda$), the input current $i(t)$ is roughly the first-order differential of input voltage $U(t)$ because the input impedance of a short dipole approximates to a pure capacitance. For broadband well-matched cases, the input current is of the same waveform as the input voltage. For other cases, the input current may have a different waveform from the input voltage, owing to the reflection at the dipole input. The varied current waveform on the dipole radiates pulse waveforms different from the original source pulse.

Next, a broadband planar dipole is considered for comparison with the aforementioned narrowband thin-wire dipole. Two 18 mm ×18 mm planar radiators are used to replace the thin-wire radiators and positioned face-to-face.[51] For comparison purposes, the source pulses used in Figure 5.15 are the same as those used in Figure 5.18. The observed fields are the co-polarized components $|E_{\theta rad}|$ in the direction $\theta = 90°$ and at distance $r = 1960$ mm, as shown in Figure 5.15. Figure 5.21 shows the power spectral density of the radiated pulses, the transfer function, and return loss $|S_{11}|$. It is readily seen that not all the radiated power spectral density can comply with the FCC's emission mask well, because the broadband dipole acting as an allpass filter with a flat transfer function $|H_{rad}(\omega)|$ hardly tailors the spectra across the bandwidth wider than the UWB band. This suggests that the broadband antenna designs are suitable for scenarios where the spectra of the source pulses conform to FCC's emission limit mask.

The Overall Transmit/Receive Antenna System

As noted earlier, one crucial criterion of UWB is associated with the overall performance of a transmit/receive antenna system. This stems from the fact that UWB systems hardly maintain invariable performance across a range of a few gigahertz. The variation in performance significantly affects the waveforms and spectra of radiated pulses, as discussed above. As a result, signal distortion at a receiver is usually severe. The transfer function can be used to assess the performance of the antenna system and evaluate the distortion of the received signals. However, it is difficult to exactly formulate the transfer function between unspecified transmit/receive antennas in a closed form owing to the complicated frequency-dependent characteristics of the antennas.[41-43] Therefore, the effects of the transfer function $H(\omega)$ given in equation 5.6 on the output signals are evaluated. Also, the performance of the antenna system is evaluated by pulse fidelity for the single-band scheme and system transmission efficiency.

The received signals, which are transmitted through both narrowband and broadband antenna systems, are compared for single-band and multi-band schemes. The source pulses used in Figures 5.16 and 5.17 are adopted, respectively. The narrowband and broadband antenna systems comprise a pair of thin-wire dipoles and planar square dipoles (see Figure 5.18).

Figure 5.22 illustrates the system transfer function $|H(\omega)|$ and the radiation transfer function $|H_{rad}(\omega)|$ for the narrowband and broadband antenna systems. This comparison reveals that the broadband antenna system features a flatter $|H(\omega)|$ or $|H_{rad}(\omega)|$ than the narrowband antenna system within the UWB band.

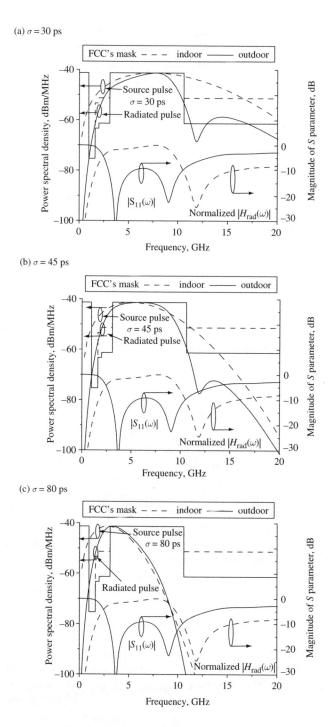

Figure 5.21 Power spectral density shaping of radiated electrical fields by a broadband planar dipole driven by first-order Rayleigh source pulses. (Reproduced by permission of IEEE.[39])

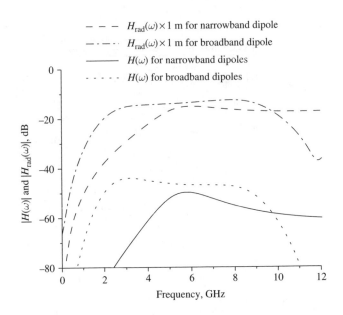

Figure 5.22 Magnitude of the system transfer function $|H(\omega)|$ and radiated transfer function $|H_{rad}(\omega)| \times 1$ m for narrowband and broadband antenna systems. (Reproduced by permission of IEEE.[39])

Figure 5.23 shows the waveforms of the received pulses in the single-band and multi-band schemes, for a narrowband antenna system (a pair of thin-wire dipoles). The pulse waveforms illustrated in (a) and (b) are not identical with the waveforms of source pulses nor the radiated pulses shown in Figure 5.19. In the single-band scheme, the severe distortion results mainly from the narrowband filtering of the antenna systems. In other words, the dispersion in the magnitude and phase of $|H(\omega)|$ results in the distortion of the pulse waveforms. In a multi-band scheme, the change in the magnitude of $|H(\omega)|$ causes the uneven envelope of the magnitudes of the received signals, which accords with the shape of the magnitude of $|H(\omega)|$ shown in Figure 5.22. The unequal amplitudes of the signals result in a different signal-to-noise ratio (S/N) in the sub-bands. Figure 5.23(c) displays the group delay and the variation in the group delay, which can distort the signals, especially when a large change in the group delay occurs. This differentiates the design considerations for UWB antenna systems from narrowband antenna systems that is able to achieve a linear phase response across the narrow operating band. The important influence of the group delay leads to shift in the carriers. This means that the maximum energy can be detected at a frequency different from the original carrier.

Similarly, a broadband antenna system (a pair of planar dipoles) is used to transmit and receive the pulses. The waveforms of the received pulses in the single-band and multi-band schemes are illustrated in Figure 5.24. Compared with the pulse waveforms illustrated in Figures 5.23(a) and (b), the pulse waveforms shown in Figures 5.24(a) and (b) exhibit a lesser variation. The pulse waveforms in the single-band scheme are similar to the second-order Rayleigh pulses. This turns on the fact that, in the time domain, the reflection occurring at the ends of the dipoles essentially brings about the distortion of the waveforms even though

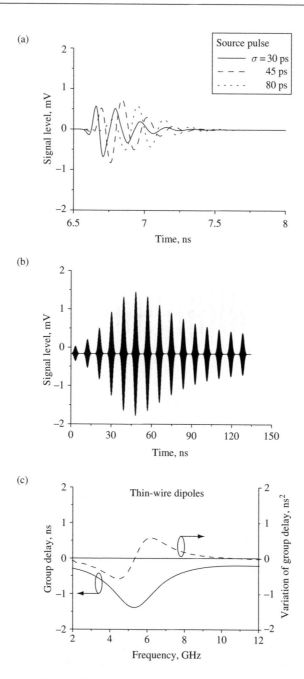

Figure 5.23 Thin-wire dipoles: (a) Waveforms of received pulses in a single-band scheme. (b) Waveforms of received pulses in a multi-band scheme. (c) The group delay and its variation. (Reproduced by permission of IEEE.[39])

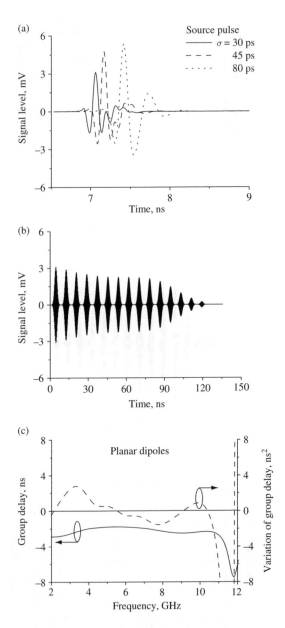

Figure 5.24 Planar dipoles: (a) Waveforms of received pulses in a single-band scheme. (b) Waveforms of received pulses in a multi-band scheme. (c) The group delay and its variation. (Reproduced by permission of IEEE.[39])

the source is well matched to the input of the antenna within a broad bandwidth. Because of the flat magnitude response of $|H(\omega)|$, the envelope of the pulse magnitudes in the multi-band scheme are more even than that shown in Figure 5.23(b), which also agrees with the shape of the $|H(\omega)|$ shown in Figure 5.22. However, it should be noted that the group delay and the variation of the group delay depicted in Figure 5.24(c) are greater than those shown in Figure 5.23(c), since the planar dipoles have larger size than the thin-wire dipoles.

Figure 5.25 demonstrates the frequency shifts of the modulated pulses for the 15 carriers. The frequency shift is closely related to the system response displayed in Figure 5.22. This suggests that in the multi-band scheme, the signals are primarily affected by the magnitude of the system response because the group delay varies slightly within one width-limited sub-band.

To evaluate the transmitting and receiving capability of the antenna system, equation 5.5 is rewritten as

$$10\lg\left(\frac{P_r(\omega)}{P_t(\omega)}\right) = \eta\,(\mathrm{dB}) - 10\log(4\pi r^2)$$

$$\eta\,(\mathrm{dB}) = 10\lg\left[\left(1 - |\Gamma_t(\omega)|^2\right)\left(1 - |\Gamma_r(\omega)|^2\right)G_t(\omega)G_r(\omega)|\hat{\rho}_t(\omega)\cdot\hat{\rho}_r(\omega)|^2\left(\frac{\lambda^2}{4\pi}\right)\right] \tag{5.12}$$

where the term η is independent of the distance between the transmitting and receiving antennas and indicates the transmit/receive capability of the antenna system.

Figure 5.26 shows the transmitting and receiving capability for the narrowband and broadband antenna systems. For both single-band and multi-band schemes, the broadband antenna system always transmits and receives the pulses much more efficiently than the narrowband antenna system. Figure 5.26(a) further suggests that, for a given antenna system in a single-band scheme, the system efficiency is dependent also on the pulse width. Moreover, Figure 5.26(b) suggests that the efficiency of a given antenna system varies with the carriers applied to a multi-band scheme.

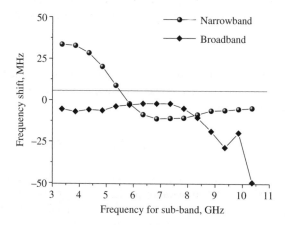

Figure 5.25 Frequency shift in terms of maximum fidelity at the varied detecting frequencies for a broadband antenna system. (Reproduced by permission of IEEE.[39])

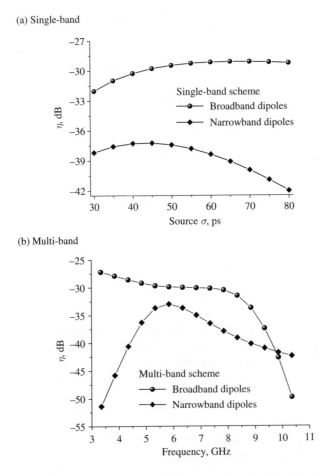

Figure 5.26 System transmission efficiency η for narrowband and broadband antenna systems: (a) single-band scheme; (b) multi-band scheme. (Reproduced by permission of IEEE.[39])

Furthermore, the fidelity of the signal of an antenna system is calculated to assess the quality of a received pulse and select a proper detection template, particularly for the single-band scheme.[40] The fidelity can be written as

$$F = \max_{\tau} \int_{-\infty}^{\infty} L[p_{source}(t)]p_{output}(t-\tau)dt \tag{5.13}$$

where $p_{source}(t)$ and $p_{output}(t)$ are the source and output pulses normalized by their energy, respectively. The fidelity F is the maximum of the integral by varying time delay τ. The linear operator $L[\cdot]$ operates on the input pulse $p_{source}(t)$. Evidently, the template at the output of a receiving antenna may be $L[p_{source}(t)]$, and not the simple $p_{source}(t)$ for maximum fidelity. The calculated F and different operators $L[\cdot]$ are tabulated in Table 5.1, where the source pulse is the first-order Rayleigh pulse. Several points can readily be observed from the table.

Table 5.1 Calculated fidelity for the pulses in the single-band scheme.

Antenna system	σ (ns) for source $p_{source(t)}$	F template p_{source}	F/ω (GHz) template $\sin(\omega t)$	$L[p_{source(t)}] = d^{n-1}/dt^{n-1}[p_{source(t)}]$		
				F	n	σ (ns)
Narrowband	30	0.74	0.80/6.16	0.89	6	83
	45	0.77	0.83/5.82	0.95	7	99
	80	0.58	0.88/5.25	0.99	12	149
Broadband	30	0.70	0.62/3.86	0.94	4	78
	45	0.81	0.72/3.73	0.93	4	91
	80	0.87	0.87/3.21	0.95	4	126

First, the waveforms of the received pulses are not identical with those of the source pulses, especially for narrowband antenna systems. Second, using sinusoidal templates, the fidelity F for the narrowband antenna system is much higher than that for the broadband one with the bandpass filtering function. Third, using optimal Rayleigh's templates, the fidelity F can be increased greatly up to more than 0.9. The order $n > 1$ of the Rayleigh pulse suggests the differential functions of the antenna system. For the broadband antenna system, the order $n = 4$ for all the three pulses. However, the order n for the narrowband antenna system is larger than those for the broadband antenna system and increases as the pulse becomes wider or the ratio of $L/\sigma c$ reduces.

Conclusions

The key considerations for source pulses and transmit/receive antenna systems for UWB in both single-band and multi-band schemes include the following:

- The radiation levels of UWB signals should comply with the emission masks. The emission characteristics can be described by a radiation transfer function. For a multi-band scheme, properly selecting the envelope of modulated source pulses can effectively control the spectrum shaping of radiated signals. For a single-band scheme, one can shape the spectra of radiated signals by properly selecting the source pulses with the spectrum conforming to the emission masks, using RF filters to tailor the spectrum shaping of radiated signals as well as the filtering function of transmitting antennas.
- The antenna system should transmit and receive signals efficiently. This can be evaluated in terms of a system transfer function. Within the UWB band, the magnitude of the transfer function should be invariable or flat, and the phase of the transfer function should be linear or the group delay kept constant so that the received signals will have little distortion.
- Antennas should be designed to be as small as possible. This is not only to reduce the size of the system but also to suppress the group delay.

- For a given antenna system, the source pulses and templates at a receiver can be optimized for high fidelity and efficiency.
- Unlike the conventional optimization of an antenna, the emission levels are taken into account in the optimization.[52] The analysis and optimization procedure can be carried out simply by a software tool.[53]

5.3.3 PLANAR UWB ANTENNA AND ASSESSMENT

The unique characteristics of UWB systems mean that the design considerations for the source pulses and antenna systems are different from those for conventional broadband antennas, as elaborated above. Now, as an example, one commonly used broadband planar antenna is optimized for acceptable transfer performance.

Recently, the effects of source and template pulses on a diamond antenna system were reported.[54,55] Planar diamond antennas are here optimized for single-band and multi-band UWB radio systems. The objective of the optimization is to obtain a flat $|S_{21}|$ response of a corresponding antenna system over the UWB band.

Three antennas, A, B and C, are shown in Figure 5.27. Each of them has been evaluated in an antenna system, where the transmitting and receiving antennas are identical and separated by a distance $R = 655$ mm. The systems will be referred to here as Antennas A/A, B/B and C/C. They have been investigated with the commercial EM simulator XFDTD in both time and frequency domains. In the simulations, each of the transmitting antennas was fed by a voltage source with an internal resistance of 100 Ω, and the loading of each of the receiving antennas is a 100-Ω resistance.

To validate the simulated results, Antenna B/B was measured with a network analyzer (HP8510c) in the frequency domain. Then, the corresponding time domain characteristics were examined with inverse FFT. Figure 5.28 shows the geometry of the antenna system under test and the measurement setup, where two equivalent monopoles of Antenna C are used. The monopoles installed vertically above the electrically large ground plane are used to approximate a pair of dipoles according to the image theory. The advantage of using monopoles in the tests is to avoid the effects of the RF cable on antenna performance. The

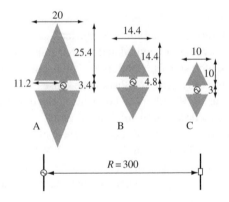

Figure 5.27 Geometry of planar diamond antennas (dimensions in millimeters). (Reproduced by permission of John Wiley & Sons, Ltd.[55])

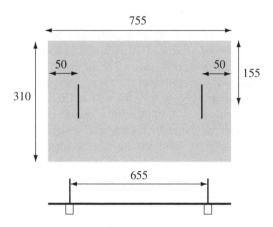

Figure 5.28 Measurement setup for a pair of planar diamond antennas – Antenna B/B (dimensions in millimeters). (Reproduced by permission of John Wiley & Sons, Ltd.[55])

transmitting and receiving antennas were mounted face-to-face above a 755 mm × 310 mm ground plane, and connected to two 50-Ω probes.

While choosing the face-to-face orientation as shown in Figure 5.28, the simulated and measured S parameters of Antenna B/B are plotted in Figure 5.29. The time-domain output pulse of an antenna system can be obtained by the inverse FFT

$$r(t) = \frac{1}{2\pi} \int_{-\infty}^{\infty} \int_{-\infty}^{\infty} S_{21}(\omega) e^{j\omega\tau} s(t-\tau) d\omega d\tau \qquad (5.14)$$

when the source pulse is chosen in the form of $g_{\sin}(t)$ as defined in

$$s(t) = \sin(2\pi f_c t) e^{\left(\frac{t}{\sigma}\right)^2} \qquad (5.15)$$

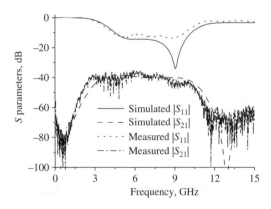

Figure 5.29 Comparison of measured and simulated S parameters of a pair of planar diamond antennas – Antenna B/B – when $\theta = \phi = \theta' = \phi' = 90°$ (face-to-face). (Reproduced by permission of John Wiley & Sons, Ltd.[55])

with $\sigma = 91$ ps and $f_c = 6.85$ GHz. Such a pulse has the -10-dB bandwidth overlapping with the UWB band. Instead of direct time-domain measurement, frequency-domain measurement with high accuracy and the use of inverse FFT is selected in the investigation. The latter has the advantage of easy calibration over the direct TD method.

The simulated and measured waveforms of the output pulses at the output loading of the receiving antenna are compared in Figure 5.30. The inconsistency between the simulated and measured pulses in the later ringing is due to the finite ground plane used in the measurement setup. In particular, edges behind the antennas within a distance of 50 mm cause reflection of the pulses. Figures 5.29 and 5.30 validated the measurement and simulation well in both time and frequency domains.

Next, the performance of the antennas is compared in Figure 5.31, and the impedance bandwidths of $|S_{11}| < -10$ dB are tabulated in Table 5.2. From the conventional broadband

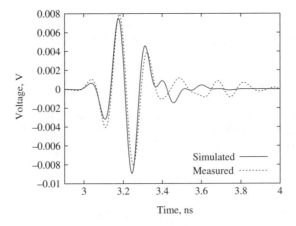

Figure 5.30 Comparison of simulated and measured waveforms at a receiver through a pair of planar diamond antennas – Antenna B/B – when $\theta = \phi = \theta' = \phi' = 90°$ (face-to-face). (Reproduced by permission of John Wiley & Sons, Ltd.[55])

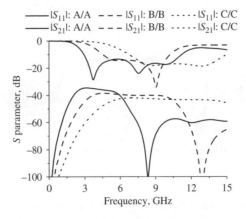

Figure 5.31 S-parameter responses of Antennas A/A, B/B and C/C when $\theta = \phi = \theta' = \phi' = 90°$ (face-to-face). (Reproduced by permission of John Wiley & Sons, Ltd.[55])

Table 5.2 The -10-dB bandwidth of $|S_{11}|$.

	Lower frequency(GHz)	Upper frequency(GHz)	Bandwidth (GHz)
A/A	2.9	11.2	8.3
B/B	4.5	10.3	5.8
C/C	6.7	15.0	8.3

Table 5.3 The -3-dB bandwidth of $|S_{21}|$.

	Lower frequency (GHz)	Upper frequency (GHz)	Bandwidth (GHz)
A/A	2.3	5.5	3.2
B/B	3.4	10.0	6.65
C/C	5.1	14.6	9.0

design, Antenna A is the best solution among the three because of its broadest impedance bandwidth covering the entire UWB band; in contrast, Antennas B and C have impedance bandwidths covering only a portion of the UWB band. However, from Figure 5.31 and Table 5.3 it is seen that Antenna B/B is the optimal design since its system bandwidth of $|S_{21}| < -3$ dB covers almost the entire UWB band, whereas the $|S_{21}|$ responses of Antennas A/A and C/C feature steep slopes in the bands 5.5–10.6 GHz and 3.1–5.1 GHz, respectively. As discussed earlier, from a systems point of view, a flat $|S_{21}|$ response over the UWB band is desirable in order to alleviate signal distortion.

The uneven $|S_{21}|$ responses of Antennas A/A and C/C are due to different reasons. For Antenna A/A, the nonlinear gain of Antenna A causes the uneven response. Although its return loss is small over the UWB band, which means most of the power in the UWB band is well radiated into free space, such radiation is not evenly distributed in all directions and at all frequencies. In the $\theta = \phi = 90°$ direction, Antenna A radiates little power at the higher frequencies of the band, namely 7.2–9.9 GHz, since $G_{abs}(\omega) < -5$ dBi as shown in Figure 5.32.

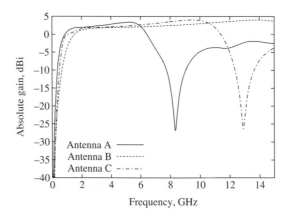

Figure 5.32 Gain for Antennas A/A, B/B and C/C when $\theta = \phi = \theta' = \phi' = 90°$ (face-to-face). (Reproduced by permission of John Wiley & Sons, Ltd.[55])

Similarly, the receiving capacity of Antenna A at these frequencies is also low in the same direction. Thus, nearly no power in such a band is received by the receiving antenna. For Antenna C/C, the change in $|S_{21}|$ is due to the varying return loss, because the gain of Antenna C changes slightly, which gradually rising from 1.75 dBi at 3.1 GHz to 3.17 dBi at 10.6 GHz as shown in Figure 5.32. Therefore, the achieved gain for Antenna B/B can best cover the UWB band.

Since it is quite difficult to keep both the return loss and absolute gain ideal over the entire UWB band simultaneously, it is necessary to make a tradeoff between the two parameters in order to achieve a relatively flat $|S_{21}|$ response. Antenna B is optimized under this consideration to achieve a relatively flat $|S_{21}|$ response by making both the return loss and gain acceptable over the UWB band.

In the single-band scheme, the source pulse is the same as that used in Figure 5.30. Figure 5.33 plots the waveforms of the received pulses for all the three antenna systems, where $\theta = \phi = \theta' = \phi' = 90°$. It can be seen that the optimal design Antenna B/B leads to a signal with the largest amplitude while keeping a narrow pulse width.

In a multi-band scheme, the source pulse is in the form described in Figure 5.16. The waveforms of the received signals are given in Figures 5.34 for Antennas A/A, B/B and C/C, where the antennas are also located face-to-face. It can be seen that the uneven $|S_{21}|$ responses severely affect the amplitudes of received pulses in different sub-bands. A relatively flat $|S_{21}|$ response leads to a good performance in (b), since the received signals in all the sub-bands can be detected well. However, some sub-bands in (a) and (c) may not be in use properly as the received power is too low for detection.

The frequency shift, which is defined as the difference between the central frequencies of the received and source pulses, is plotted in Figure 5.35 for each sub-band. Generally, a flat $|S_{21}|$ leads to a small frequency shift. Therefore, a relatively flat $|S_{21}|$ response is required for a multi-band scheme, in order to achieve similar magnitude responses for different sub-bands, and to minimize the frequency shift.

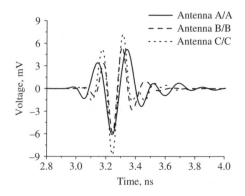

Figure 5.33 Comparison of waveforms of the received pulses for Antennas A/A, B/B and C/C under a single-band scheme when $\theta = \phi = \theta' = \phi' = 90°$ (face-to-face). (Reproduced by permission of John Wiley & Sons, Ltd.[55])

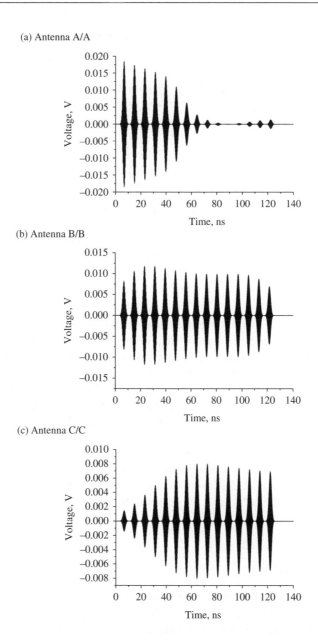

(a) Antenna A/A

(b) Antenna B/B

(c) Antenna C/C

Figure 5.34 Comparison of waveforms of the received pulses for Antennas A/A, B/B and C/C under a multi-band scheme when $\theta = \phi = \theta' = \phi' = 90°$ (face-to-face). (Reproduced by permission of John Wiley & Sons, Ltd.[55])

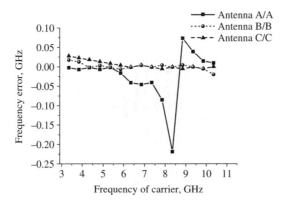

Figure 5.35 Frequency shift for Antennas A/A, B/B and C/C under a multi-band scheme when $\theta = \phi = \theta' = \phi' = 90°$ (face-to-face). (Reproduced by permission of John Wiley & Sons, Ltd.[55])

By changing the orientation of the transmitting and receiving antennas, the $|S_{21}|$ responses in other directions were examined and plotted in Figures 5.36 for Antennas A/A, B/B and C/C. The orientations of the two antennas are changed in the same way, namely $\theta = \theta_A = \theta_{A'}$ and $\phi = \phi_A = \phi_{A'}$ are assumed. Two observations can be made from the results. First, the $|S_{21}|$ responses of all the systems feature relatively smaller variation in the $\theta = 90°$ plane than those in the $\phi = 90°$ plane. Second, compared with Antenna A/A and C/C, Antenna B/B shows the flattest $|S_{21}|$ responses over the UWB band in all the directions in the $\theta = 90°$ plane.

Therefore, it can be concluded that the relatively flat $|S_{21}|$ response is the key to achieving acceptable performance of the antenna system, and to achieve a flat $|S_{21}|$ response, a tradeoff between the impedance matching at lower operating frequencies and the gain at higher operating frequencies should be made for antennas that may not cover the whole UWB band well.

5.4 CASE STUDIES

There have been many candidates for a variety of UWB applications. The basic requirements include a broad frequency response for impedance matching, gain, polarization and phase, which can be evaluated in both time and frequency domains. In practice, the situation becomes much more complicated, especially the effects of the environment where the antennas are installed; should be taken into account.

In this section of the chapter, three design cases are introduced. The first example is of a planar square dipole printed on a PCB with a system ground; the effects of the system ground plane and dielectric substrate on performance are addressed. Then, a small planar monopole embedded into a laptop computer is designed for potential WLAN applications. Last, a directional planar Vivaldi antenna is presented for radar applications.

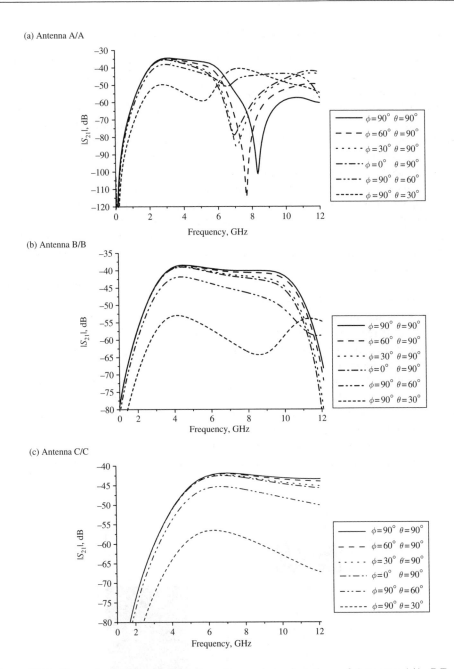

Figure 5.36 Comparison of transfer functions for varying orientations of Antennas A/A, B/B and C/C. (Reproduced by permission of John Wiley & Sons, Ltd.[55])

5.4.1 PLANAR UWB ANTENNA PRINTED ON A PCB

In portable devices, the antenna is often embedded into the casing, or integrated on to the printed-circuit board with other RF circuits. The effects of the PCB and system ground plane on the performance should be taken into account.[56]

Figure 5.37 shows the geometry of a balance-fed antenna with two square radiators of size 12 mm $\times$ 12 mm, which is printed on a 70 mm $\times$ 100 mm PCB having a thickness $h = 1.5$ mm and dielectric constant $\epsilon_r = 3.38$. The 40 mm $\times$ 100 mm system ground plane is placed at a distance $d = 13$ mm from the dipole bottom. The gap between the two radiators is 7 mm. A 100-Ω balanced source excites the squares at their midpoints.

In tests, the transmit/receive antenna system consists of a pair of identical square balance-fed monopoles (namely, half of planar dipoles), which are positioned face-to-face with a separation of 655 mm as shown in Figure 5.38. The transmitting and receiving antennas are mounted above a 755 mm $\times$ 310 mm ground plane and connected to two 50-Ω probes. A network analyzer (Agilent E8364B) is used to measure the S parameters of the antennas. An XFDTD simulator was used.

From Figure 5.39, it can be seen that there is good impedance matching of lower than -10 dB, ranging from 3 GHz to 9.5 GHz. The relatively flat transfer function $|S_{21}|$ covers the passband of the antenna system. As mentioned above, a flat transfer function will limit the distortion of received signals. However, at higher frequencies of the passband (above 8 GHz), the transfer function drops rapidly owing to the significant change in the gain in the measurement direction (face-to-face).

Figure 5.40 depicts the waveform of the received pulse by the planar dipoles driven by a monocycle pulse shown in Figure 5.17, with $\sigma = 65$ ps using a single-band scheme. The waveform of the received pulse has more ringing than the source pulse. However, the distortion is not severe owing to the flat transfer function.

Figure 5.41 illustrates the waveforms of the received modulated pulses shown in Figure 5.16 in a multi-band scheme. The waveforms are distorted owing to the lower $|S_{21}|$

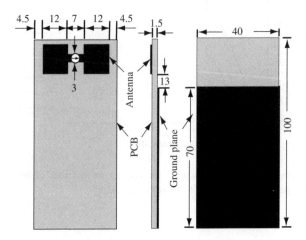

Figure 5.37 Geometry of a planar square dipole printed on a PCB. (dimensions in millimeters)

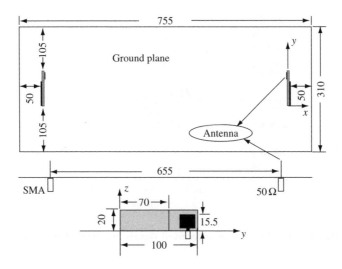

Figure 5.38 Measurement setup for a pair of planar square dipoles printed on PCBs (dimensions in millimeters).

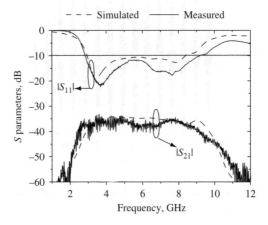

Figure 5.39 Simulated and measured S parameters of planar square balance-fed antennas printed on a finite PCB and ground plane.

response at higher frequencies. The change in the magnitude of the received pulses agrees with the measured $|S_{21}|$ response.

Figure 5.42 shows the simulated radiation patterns at 3 GHz and 8 GHz in three main planes. In x–z plane, the radiation patterns remain almost the same at 3 GHz and 8 GHz. As against the omnidirectional radiation patterns of dipoles in free space, the dipoles on the PCB have directional radiation due to the system ground plane. In y–z and x–y planes, the radiation patterns change significantly for varying frequencies. The system ground plane causes severe degradation, in particular at 8 GHz. Therefore, the effect of the system ground

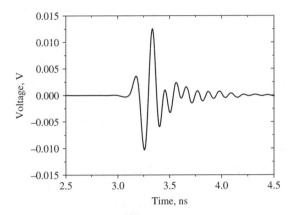

Figure 5.40 Waveforms of received signals in a single-band scheme.

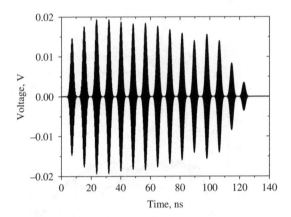

Figure 5.41 Waveforms of received signals in a multi-band scheme.

plane on the performance of antennas printed on the PCB is significant. In the following discussion, the size of the square monopole is 15 mm × 15 mm and the gap between the dipoles is 5.5 mm. The distance between two test antennas is 300 mm. For comparison, the results for the dipole in free space are provided simultaneously.

Figures 5.43–5.45 show the effects of the system ground plane on performance. The distance from the ground plane to the dipoles' bottom is varied and the thickness is 1.5 mm. Figure 5.43 shows that the system ground plane affects the S parameter response at higher frequencies when $d = 3$ mm. As a result, the bandwidth becomes narrower as the impedance matching worsens, and the gain decreases rapidly beyond 7 GHz. When $d = 15$ mm, the gain increases a little due to the better impedance matching. However, the transfer function $|S_{21}|$ becomes uneven.

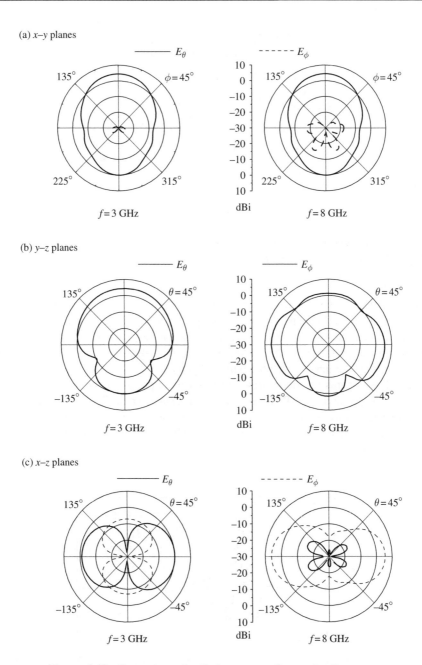

Figure 5.42   Comparison of radiation patterns for varying frequency.

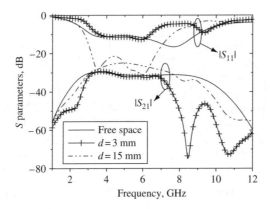

Figure 5.43 S parameters of planar dipoles printed on a PCB for varying system ground planes. (Reproduced by permission of IEEE.[56])

Figure 5.44 shows the waveforms of the received pulses in a single-band scheme with $d = 3$ mm and 15 mm. Compared with the free-space condition, the waveform for $d = 3$ mm has smaller peak values and more ringing owing to the narrower bandwidth. Owing to the higher gain, the case with $d = 15$ mm has the highest peak values. It should be noted that the dielectric substrate has delayed the pulses slightly.

Figure 5.45 compares the waveforms of the received pulses in a multi-band scheme for dipole in free space and on a PCB with $d = 3$ mm and 15 mm. The system ground plane significantly suppresses the pulses after 70 ns for the case with $d = 3$ mm and after 100 ns for $d = 15$ mm. This phenomenon accords well with what has been observed in the frequency domain in Figure 5.43.

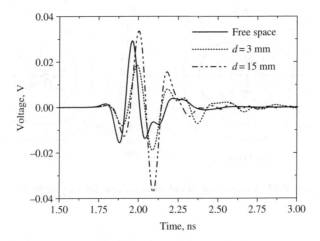

Figure 5.44 Waveforms of received pulses in a single-band scheme for varying system ground planes. (Reproduced by permission of IEEE.[56])

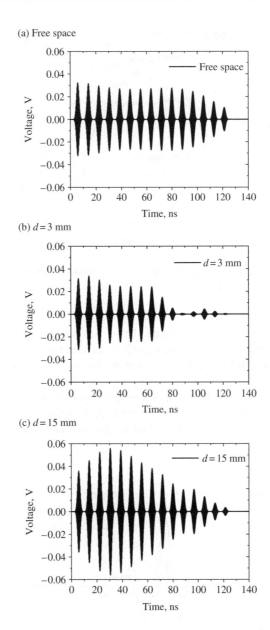

(a) Free space

(b) $d = 3$ mm

(c) $d = 15$ mm

Figure 5.45 Waveforms of received signals in a multi-band scheme for varying system ground planes. (Reproduced by permission of IEEE.[56])

Figures 5.46–5.48 show the effects of varying PCB thickness h on the antenna performance. The system ground plane is placed at a distance $d = 3$ mm from the dipoles' bottom. Figure 5.46 demonstrates that the bandwidths of the impedance and transfer functions of the dipoles on the PCBs have become much narrower because of the increase in Q

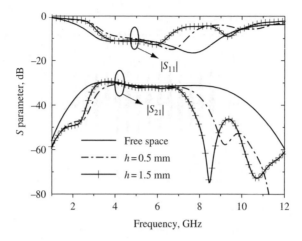

Figure 5.46 S parameters of planar dipoles printed on a PCB for varying dielectric substrate thickness. (Reproduced by permission of IEEE.[56])

of the dipoles. For example, the 3-dB bandwidth of the transfer function for the dipole with $h = 1.5$ mm is just 2.96 GHz, as against 6.56 GHz for the dipole in free space. Therefore, the thicker the dielectric substrate, the narrower the bandwidths of impedance and transfer functions.

Figure 5.47 shows the waveforms of the received pulses in a single-band scheme for varying dielectric thickness $h = 0.5$ mm and 1.5 mm. A significant effect of the dielectric substrate is the time delay of the pulses. The thicker the substrate, the larger is the time delay of the pulses.

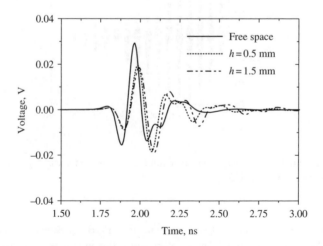

Figure 5.47 Waveforms of received pulses in a single-band scheme for varying dielectric substrate thickness. (Reproduced by permission of IEEE.[56])

The waveforms of the received pulses with varying thickness $h = 0.5$ mm and 1.5 mm in a multi-band scheme are shown in Figure 5.48. Compared with the waveforms for the dipole in free space shown in Figure 5.45(a), the thicker dielectric substrate causes a severe degradation in the waveforms, in particular at higher frequencies.

In conclusion, the presence of the dielectric substrate and the system ground plane causes significant effects on the performance of these antennas printed on PCBs, in terms of degradation of the impedance and transfer functions. As a result, the waveforms of the received pulses are distorted.

5.4.2 PLANAR UWB ANTENNA EMBEDDED INTO A LAPTOP COMPUTER

With the increasing deployment of WLANs, many laptop computers are networked wirelessly. Currently, technology based on standards such as IEEE 802.11a/b/g can provide data rates from 11 Mbps up to 108 Mbps. They operate in the frequency bands of 2.4–2.4835 GHz, 5.150–5.250 GHz, 5.250–5.350 GHz, and 5.725–5.825 GHz. Many antennas

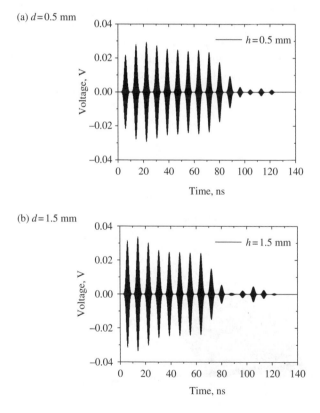

Figure 5.48 Waveforms of received pulses in a multi-band scheme for varying system ground planes. (Reproduced by permission of IEEE.[56])

for WLAN devices embedded into laptop computers have been developed. [57–60] The UWB may be a promising technology for next-generation wireless communications, especially for high data rates and short-range WLAN or WPAN applications.

One of the challenges in UWB systems is to design very small antennas covering the ultra-wide bandwidth of 3.1–10.6 GHz and having shapes that conform with portable devices. Planar designs have been considered as suitable candidates owing to their simple mechanical structures but broadband characteristics, as noted in section 5.3. However, their size is still too large for laptop computers, where very thin and low-profile antennas are strictly required. One reported design[61] has very good impedance matching within a 7:1 frequency range, but the antenna's height (>25 mm) is too high for laptop computer applications within the UWB band of 3.1–10.6 GHz.

Another design challenge is the significant proximity effect of a lossy screen display. These antennas are usually embedded into the display of a laptop computer. The lossy metal objects around the antenna not only affect the impedance matching but also absorb the energy provided by an RF cable.

This section of the chapter introduces a planar UWB antenna for potential laptop computer applications. The antenna is formed in the shape of a planar half-circle with a top rectangle, and etched on to a piece of thin PCB, as in Figure 5.49. The antenna has a low profile and can be embedded easily into the display of a laptop computer. Its simple structure and cheap material keep the cost down. The impedance and radiation performance of this antenna integrated into a lossy display of a laptop are investigated. The VSWR, maximum and average gain as well as radiation patterns are examined experimentally.

The design is suitable also for other portable devices in UWB applications. However, the effect of the environment where the antenna is installed should be carefully re-examined in each case.

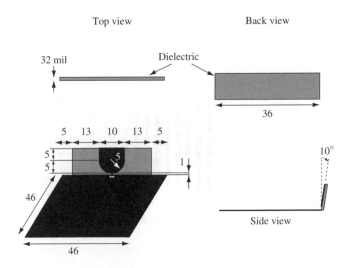

Figure 5.49 Geometry of a planar UWB antenna for laptop computer applications (dimensions in millimeters).

Consider the planar antenna shown in Figure 5.49. It consists of semicircle with radius 5 mm and a top rectangle with dimensions 5 mm ×10 mm. The antenna was etched centrally on a piece of thin PCB (Roger4003, $\epsilon_r = 3.38$) with dimensions 36 mm ×11 mm ×32 mil (about 0.81mm). All dimensions were obtained by simulation of the antenna in free space, and optimized experimentally for the antenna embedded into the lossy display of a laptop computer.

The PCB was electrically connected to a copper sheet of size 46 mm × 46 mm, which was used as a ground plane to ease the change in the performance of the antenna installed in a varying environment. The PCB tilts at an angle of 10° from the z-axis to fit the shape of the cover edge.

A 50-Ω coaxial cable excites the bottom of the hemi-circle with a feed gap of 1 mm through the ground plane to reduce the random effect of the feeding cable. By changing the feed gap, the impedance matching can be adjusted. Usually, the feed gap is around 1 mm for the UWB band 3.1–10.6 GHz. Thus, the overall height of the antenna is less than 11 mm.

To reduce the proximity effects of a human body or support surface, antennas are mainly placed in the display frame of a laptop computer. It is thus placed very close to the lossy liquid crystal display (LCD) panel. To be embedded into the frame, it needs to be low-profile and thin. In this instance the antenna was placed at the top edge of the display frame as shown in Figure 5.50. The LCD panel is lossy and usually leads to severe undesired ohmic losses. To keep the radiation effective, the top edge of the display frame has a slot 50 mm × 8 mm .

The antenna was installed centrally in the slot as shown in Figure 5.51. A lossy conducting LCD panel with thickness 6 mm lies very close to the antenna, with the least separation being 3 mm at the feed point at the bottom of the antenna. The proximity of the conducting LCD affects impedance matching to some degree, particularly at low frequencies. The ground plane of the antenna is under the LCD panel, and electrically connected to the cover. Thus, the lossy cover causes a loss, too. Investigations have shown that the LCD panel significantly affects both impedance matching and efficiency of the antenna.

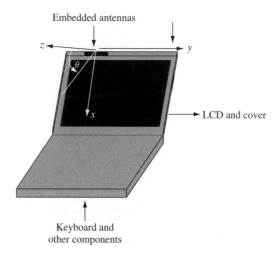

Figure 5.50 Installation position of the antenna.

Figure 5.51 Antenna installed in the frame of the laptop's cover.

The input impedance of the antenna was first measured. Compared with the antenna in free space, the embedded antenna has a broader bandwidth. It is inferred that the lossy cover and LCD panel reduce the Q of the antenna system so that a broad well-matched bandwidth can be attained. However, such losses will reduce the radiation efficiency and to some degree be harmful to antenna performance. Figure 5.52 shows the measured VSWR for the embedded antenna. The antenna has good impedance matching for VSWR $= 2.5$ across the entire UWB band from 3.1 GHz to 10.6 GHz. The result also implies that many resonances have been excited well within the UWB band.

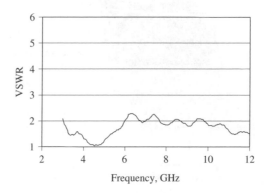

Figure 5.52 Measured VSWR of the antenna installed in the frame of the laptop's cover.

(a) Average gain

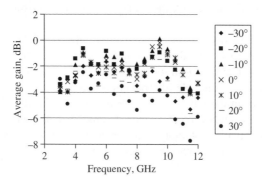

(b) Maximum gain

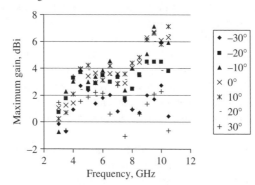

Figure 5.53 Measured maximum gain and average gain for the antenna embedded into the laptop's cover.

Figure 5.53 exhibits measured maximum and average total gain against frequency ranging from 3 GHz to 12 GHz; this includes the contributions of both vertical and horizontal polarized components. The measurement range spans the elevation planes of $\theta = -30°$ to $+30°$ with an increment of 10°.

Figure 5.53(a) reveals that, compared with the maximum gain shown in Figure 5.53(b), the average gain of the antenna varies in a lesser range from -4 dBi to 0 dBi over the UWB band. However, the average gain in the cut of $\theta = 30°$ is somewhat lower than those in other planes. From the measured gain, it is seen that the losses due to the cover and LCD panel have lowered the radiation efficiency.

Figure 5.53(b) shows that, around 3 GHz, the antenna has a lower gain of around 0 dBi. However, at the higher frequencies, the maximum gain increases up to 7 dBi. At most of the frequencies within the UWB band, the maximum gain varies from 2 dBi to 5 dBi. Such an increase in the gain with frequency is conducive to compensation for frequency-dependent path losses in system design. In the cuts of $\theta = -10°$, $0°$ and $+10°$, the antenna has shown higher gain than in other cuts. In the planes, for example, of $\theta = -30°$ and $+30°$, due to the shadow effect of the cover and display, the gain greatly decreases, though a conventional planar monopole in free space has higher radiation levels in these planes. Clearly, the cover and LCD panel significantly block the radiation from the antenna and change the radiation patterns.

Finally, Figure 5.54 displays a comparison of the far-field radiation patterns for the total gain in a cut of $\phi = 90°$ (the horizontal or y–z plane) at 3.0, 7.0 and 10.0 GHz (see Figure 5.50). The results demonstrate that the planar antenna is basically a variation of a monopole, and features monopole-like radiation patterns, which are quite consistent across the UWB band. Dips in the radiation patterns are observed in the direction of the computer user (z-direction). The higher the frequency, the higher is the gain.

There are other potential solutions, some of which are shown in Figure 5.55. Basically, they are all planar and with a smooth shape near feed points.

5.4.3 PLANAR DIRECTIONAL UWB ANTENNA

Directional UWB antennas have an important role in applications such as radar, base stations and measurements. Compared with three-dimensional antenna designs (such as TEM horns, bi-conical or discone antennas and log-periodic antennas), the planar Vivaldi antenna can offer more promise for UWB applications. The Vivaldi typically has an exponentially tapered slot, and has been widely applied in radar, wireless communications and measurement applications since 1979.[62] The initial designs were balanced structures fed by a broadband balun transformer. This led to a large size and high cost due to the broadband balun. Recent designs have replaced the balun with a simple transition and direct feed, as shown in Figure 5.56.

Here, an antipodal Vivaldi antenna will be examined, as shown in Figure 5.57.[63] Instead of the usual exponentially tapered slot, elliptical shapes are used to form the feed line transition and the tapered slot. A stripline and an elliptical tapered ground plane form a short

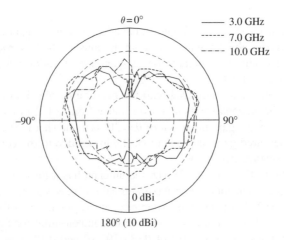

Figure 5.54 Measured radiation patterns of the antenna installed in the frame of the laptop's cover in the horizontal plane.

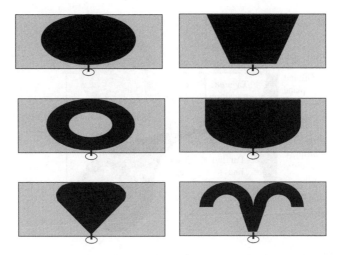

Figure 5.55 Other possible solutions for a planar antenna embedded into a laptop computer.

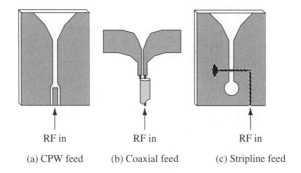

RF in	RF in	RF in
(a) CPW feed	(b) Coaxial feed	(c) Stripline feed

Figure 5.56 Geometry of planar Vivaldi antennas without balun.

transition structure to produce good performance. The transition is used to transform the 50-Ω stripline to a parallel line to excite the tapered slot.

The stripline and the elliptical tapered ground plane were etched on a PCB separated by a dielectric substrate (Roger4003) with thickness 0.8128 mm, and $\epsilon_r = 3.38 - j0.002$. The stripline is 1.86 mm in width and the parallel line 1.0 mm. The elliptical tapered transition has a major to minor axis ratio of 1.084. The conducting arms on each side of the substrate are flared in opposite directions to form the tapered slot, where the inner and outer edges of the conducting arms follow the outline of ellipses with different major to minor axis ratios. The elliptical inner edges of the conducting arms have an axis ration of 2, whereas the outer edges have a ratio of 1.2. Two semicircles are added to the ends to suppress possible diffraction. This increases the gain and lowers side-lobe levels.

Figure 5.58 shows the simulated and measured return losses of this antipodal Vivaldi antenna. The bandwidth for -10 dB return loss ranges from 2.3 GHz to 28.1 GHz.

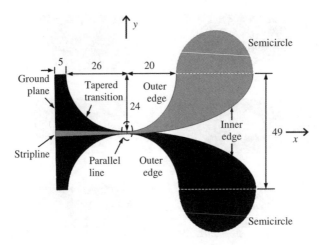

Figure 5.57 Geometry of an antipodal Vivaldi antenna (dimensions in millimeters).

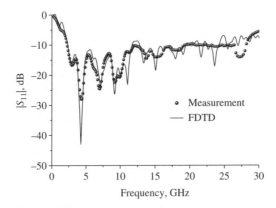

Figure 5.58 Simulated and measured return loss for the Vivaldi.

Figure 5.59 demonstrates the good agreement between the simulated and measured maximum gain in the boresight ($\theta = 90°$, $\phi = 0°$). The measured gain varies from 3.9 dBi at 3.1 GHz to 6.0 dBi at 11.6 GHz. The measured radiation patterns for E_ϕ components in both E-planes and H-planes are shown in Figure 5.60. These radiation patterns were measured at the center frequency of the UWB band, 7 GHz. Good symmetrical patterns with half-power beamwidths of 40° in the E-plane and 67° in the H-plane are achieved, which is important for radar applications. Figure 5.61 shows the measured half-power beamwidth against frequency. In general, with increasing frequency, the half-power beamwidth becomes narrower.

For UWB antenna systems, the *group delay* is a useful parameter to measure the variation of the phase response against frequency. The group delay of an antenna can be calculated from the derivative of the phase response of the transfer function with respect to frequency.

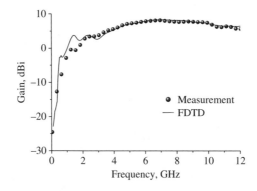

Figure 5.59 Comparison of simulated and measured gain for the Vivaldi.

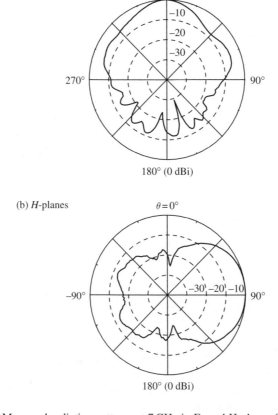

Figure 5.60 Measured radiation patterns at 7 GHz in E- and H-planes for the Vivaldi.

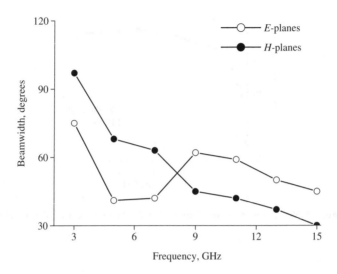

Figure 5.61 Comparison of simulated and measured gain for the Vivaldi.

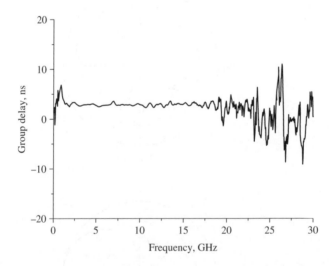

Figure 5.62 Measured group delay for the transmit and receive transfer functions in the boresight ($\theta = 90°, \phi = 0°$) for the Vivaldi.

Figure 5.62 shows the measured group delay for the Vivaldi antenna in the boresight. From 2.0 GHz to 19.0 GHz, the group delay is about 2.5 ns with little variation. The group delay sharply changes at lower and higher frequencies because of the changing nonlinear phase response.

REFERENCES

[1] H. Kawakami and G. Sato, 'Broadband characteristics of rotationally symmetric antennas and thin wire constructs,' *IEEE Transactions on Antennas and Propagation*, vol. 35, no. 1, pp. 26–32, 1987.

[2] H. Nakano, N. Ikeda, Y. Wu, R. Suzuki, H. Mimaki and J. Yamauchi, 'Realization of dual-frequency and wide-band VSWR performances using normal-mode helical and inverted-F antennas,' *IEEE Transactions on Antennas and Propagation*, vol. 46, no. 6, pp. 788–793, 1998.

[3] S. D. Rogers and C. M. Butler, 'Cage antennas optimized for bandwidth,' *Electronics Letters*, vol. 36, no. 11, pp. 932–933, 2000.

[4] W. Cho, M. Kanda, H. Hwang and M. Howard, 'A disk-loaded thick cylindrical dipole antenna for validation of an EMC test site from 30 to 300 MHz,' *IEEE Transactions on Electromagnetic Compatibility*, vol. 42, no. 2, pp. 172–180, 2000.

[5] G. Dubost and S. Zisler, *Antennas a large bande*. 1976.

[6] G. H. Brown and O. M., 'Experimentally determined radiation characteristics of conical and triangular antennas,' *RCA Review*, vol. 13, no. 4, pp. 425–452, 1952.

[7] S. Honda, M. Ito, H. Seki and Y. Jinbo, 'A disk monopole antenna with 1:8 impedance bandwidth and omnidirectional radiation pattern,' *International Symposium on Antennas and Propagation*, pp. 1145–1148, 1992.

[8] N. P. Agrawall, G. Kumar and K. P. Ray, 'Wide-band planar monopole antenna,' *IEEE Transactions on Antennas and Propagation*, vol. 46, no. 2, pp. 294 –295, 1998.

[9] M. J. Ammann, 'Square planar monopole antenna,' *IEE National Conference on Antennas and Propagation*, vol. 1, pp. 37–40, 1999.

[10] Z. N. Chen, 'Impedance characteristics of planar bow-tie-like monopole antennas,' *Electronics Letters*, vol. 36, no. 13, pp. 1100–1101, 2000.

[11] M. J. Ammann and Z. N. Chen, 'Wideband monopole antennas for multi-band wireless systems,' *IEEE Antennas and Propagation Magazine*, vol. 45, no. 2, pp. 146–150, 2003.

[12] C. Puente and R. Pous, 'Fractal design of multiband and low side-lobe arrays,' *IEEE Transactions on Antennas and Propagation*, vol. 44, no. 5, pp. 1–10, 1996.

[13] M. J. Ammann, 'Wideband antenna for mobile wireless terminals,' *Microwave and Optical Technology Letters*, vol. 26, no. 6, pp. 360–362, 2000.

[14] L. S. Lee, P. Hall and P. Gardner, 'Compact wideband planar monopole antenna,' *Electronics Letters*, vol. 35, pp. 2157–2158, 1999.

[15] M. J. Ammann, 'The pentagonal planar monopole for digital mobile terminals; bandwidth considerations and modelling,' *11th IEE International Conference on Antennas and Propagation*, vol. 1, pp. 82–85, 17–20 April 2001.

[16] Z. N. Chen, 'Experimental on input impedance of tilted planar monopole antennas,' *Microwave and Optical Technology Letters*, vol. 26, no. 3, pp. 202–204, 2000.

[17] Z. N. Chen and M. Y. W. Chia, 'Impedance characteristics of trapezoidal planar monopole antenna,' *Microwave and Optical Technology Letters*, vol. 27, no. 2, pp. 120–122, 2000.

[18] Z. N. Chen and M. Y. W. Chia, 'Broadband monopole antenna with parasitic planar element,' *Microwave and Optical Technology Letters*, vol. 27, no. 3, pp. 209–210, 2000.

[19] Z. N. Chen, 'Broadband planar monopole antenna,' *IEE Proceeedings: Microwave, Antennas and Propagation*, vol. 147, no. 6, pp. 526–528, 2000.

[20] Z. N. Chen and M. Y. W. Chia, 'Impedance characteristics of EMC triangular planar monopoles,' *Electronics Letters*, vol. 37, no. 21, pp. 1271–1272, 2001.

[21] M. J. Ammann and Z. N. Chen, 'A wideband shorted planar monopole with bevel,' *IEEE Transactions on Antennas and Propagation*, vol. 51, no. 4, pp. 901–903, 2003.

[22] Z. N. Chen, M. J. Ammann, M. Y. W. Chia and S. P. See, 'Circular annular planar monopoles with EM coupling,' *IEE Proceedings: Microwave, Antennas and Propagation*, vol. 150, no. 4, pp. 269–273, 2003.

[23] M. J. Ammann, Z. N. Chen and M. Y. W. Chia, 'Broadband square annular planar monopoles,' *Microwave and Optical Technology Letters*, vol. 36, no. 6, pp. 449–454, 2003.

[24] C. Balanis, *Antenna Theory: Analysis and Design*. New York: John Wiley & Sons, Inc., 1997.

[25] R. W. P. King, 'Asymmetric driven antenna and the sleeve dipole,' *Proceedings of the IRE*, vol. 38, no. 12, pp. 1154 –1164, 1950.

[26] Z. N. Chen, K. Hirasawa and K. Wu, 'A novel top-sleeve monopole in two parallel plates,' *IEEE Transactions on Antennas and Propagation*, vol. 49, no. 3, pp. 438–443, 2001.

[27] Z. N. Chen, 'Broadband roll monopole,' *IEEE Transactions on Antennas and Propagation*, vol. 51, no. 11, pp. 3175–3177, 2003.

[28] Z. N. Chen, M. Y. W. Chia and M. J. Ammann, 'Optimization and comparison of broadband monopoles,' *IEE Proceeedings: Microwave, Antennas and Propagation*, vol. 150, no. 6, pp. 429–435, 2003.

[29] *Instruction Manual: Type S-2 Sampling Head*. S.W. Millikan Way, P.O. Box 500, Beaverton, Oregon 97005, USA: Tektronix, Inc., 1968.

[30] G. F. Ross, *Transmission and reception system for generating and receiving baseband duration pulse signals for short base-band pulse communication system*. US Patent 3,728,632, 1973.

[31] C. L. Bennett and G. F. Ross, 'Time-domain electromagnetics and its applications,' *Proceedings of the IEEE*, vol. 66, no. 3, pp. 299–318, 1978.

[32] L. Carin and L. B. Felsen, *Ultra-wideband Short-pulse Electromagnetics, 2*, New York: Plenum, 1995.

[33] H. F. Harmuth, *Nonsinusoidal Waves for Radar and Radio Communication*. New York: Academic Press, 1981.

[34] *First Report and Order*. 445, 12th Street, S.W. Washington, DC. 20554, USA: Federal Communications Commission (FCC), 14 February 2002.

[35] M. Z. Win, R. A. Scholtz and M. A. Barnes, 'Ultra-wide bandwidth signal propagation for indoor wireless communications,' *IEEE International Conference on Communications*, vol. 1, pp. 56–60, June 1997.

[36] R. A. Scholtz, R. J. Cramer and M. Z. Win, 'Evaluation of an ultra-wide-band propagation channel,' *IEEE Transactions on Antennas and Propagation*, vol. 50, no. 5, pp. 561–570, 2002.

[37] M. Welborn and J. McCorkle, 'The importance of fractional bandwidth in ultra-wideband pulse design,' *IEEE International Conference on Communications*, vol. 2, pp. 753–757, June 2002.

[38] M. Z. Win and R. A. Scholtz, 'On the robustness of ultra-wide bandwidth signals in dense multipath environments,' *IEEE Communications Letters*, vol. 2, no. 2, pp. 51–53, 1998.

[39] Z. N. Chen, X. H. Wu, N. Y. H. F. Li and M. Y. W. Chia, 'Considerations for source pulses and antennas in UWB radio systems,' *IEEE Transactions on Antennas and Propagation*, vol. 52, no. 7, pp. 1739–1748, 2004.

[40] D. Lamensdorf and L. Susman, 'Baseband-pulse-antenna techniques,' *IEEE Antennas and Propagation Magazine*, vol. 36, no. 1, pp. 20–30, 1994.

[41] G. F. Ross, 'A time-domain criterion for the design of wideband radiating elements,' *IEEE Transactions on Antennas and Propagation*, vol. 16, no. 3, p. 355, 1968.

[42] A. Shlivinski, E. Heyman and R. Kastner, 'Antenna characterization in the time domain,' *IEEE Transactions on Antennas and Propagation*, vol. 45, no. 7, pp. 1140–1149, 1997.

[43] D. M. Pozar, 'Waveform optimizations for ultra-wideband radio systems,' *IEEE Transactions on Antennas and Propagation*, vol. 51, no. 9, pp. 2335–2345, 2003.

[44] H. J. Schmitt, J. C. W. Harrison and J. C. S. Williams, 'Calculated and experimental response of thin cylindrical antenna to pulse excitation,' *IEEE Transactions on Antennas and Propagation*, vol. 14, no. 2, pp. 120 – 123, 1966.

[45] R. J. Palciauskas and R. E. Beam, 'Transient fields of thin cylindrical antennas,' *IEEE Transactions on Antennas and Propagation*, vol. 18, no. 2, pp. 276–278, 1970.

[46] D. M. Pozar, R. E. McIntosh and S. G. Walker, 'The optimum feed voltage for a dipole antenna for pulse radiation,' *IEEE Transactions on Antennas and Propagation*, vol. 31, no. 4, pp. 563–569, 1983.

[47] D. M. Pozar, D. H. Schaubert and R. E. McIntosh, 'The optimum transient radiation from an arbitrary antenna,' *IEEE Transactions on Antennas and Propagation*, vol. 32, no. 7, pp. 633–634, 1984.

[48] M. Onder and M. Kuzuoglu, 'Optimal control of the feed voltage of a dipole antenna,' *IEEE Transactions on Antennas and Propagation*, vol. 40, no. 4, pp. 414–421, 1992.

[49] Z. N. Chen, X. H. Wu, N. Yang and M. Y. W. Chia, 'Design considerations for antennas in UWB wireless communication systems,' *IEEE International Symposium on Antennas and Propagation*, vol. 1, pp. 822–825, June 2003.

[50] G. S. Smith, 'On the interpretation for radiation from simple current distributions,' *IEEE Antennas and Propagation Magazine*, vol. 40, no. 3, pp. 9–14, 1998.

[51] Z. N. Chen, X. H. Wu and N. Yang, 'Broadband planar monopoles for UWB radio systems,' *IEEE Transactions on Antennas and Propagation*, vol. 53, no. 7, pp. 2178–2184, July 2005.

[52] D. M. Shan, Z. N. Chen and X. H. Wu, 'Signal optimization for UWB radio systems,' *IEEE Asia-Pacific Microwave Conference*, vol. 3, pp. 1977–1980, November 2003.

[53] X. H. Wu and Z. N. Chen, 'Design and optimization of UWB antennas by a powerful CAD tool: PULSE KIT,' *IEEE International Symposium on Antennas and Propagation*, vol. 2, pp. 1756–1759, June 2004.

[54] H. G. Schantz and L. Fullerton, 'The diamond dipole: a Gaussian impulse antenna,' *IEEE International Symposium on Antennas and Propagation*, vol. 4, pp. 100–103, 8–13 July 2001.

[55] X. H. Wu, Z. N. Chen and N. Yang, 'Optimization of planar dimond antenna for single/multi-band UWB wireless communications', *Microwave and Optical Technology Letters*, vol. 42, no. 6, pp. 451–455, Sept. 2004.

[56] Y. Zhang, Z. N. Chen and M. Y. W. Chia, 'Characteristics of planar dipoles printed on finite-size PCBs in UWB radio systems,' *IEEE International Symposium on Antennas and Propagation*, vol. 3, pp. 2512–2515, June 2004.

[57] R. R. Ramirez and F. DeFlaviis, 'Triangular microstrip patch antennas for dual-mode 802.11a,b WLAN applications,' *IEEE International Symposium on Antennas and Propagation*, vol. 4, pp. 44–47, June 2002.

[58] S. Yeh and K. L. Wong, 'Dual-band F-shaped monopole antenna for 2.4/5.2-GHz WLAN application,' *IEEE International Symposium on Antennas and Propagation*, vol. 4, pp. 72–75, 2002.

[59] D. Liu, B. Gaucher, E. Flint, T. Studwell and H. Usui, 'Developing integrated antenna subsystems for laptop computers,' *IBM Journal of Research & Development*, vol. 47, no. 2/3, pp. 355–367, 2003.

[60] D. Liu and B. Gaucher, 'A triband antenna for WLAN applications,' *IEEE International Symposium on Antennas and Propagation*, vol. 2, pp. 18–21, June 2003.

[61] S. Y. Suh, W. L. Stutzman and W. A. Davis, 'A new ultra-wideband printed monopole antenna: the planar inverted cone antenna (PICA),' *IEEE Transactions on Antennas and Propagation*, vol. 52, no. 5, pp. 1361–1364, 2004.

[62] P. Gibson, 'The Vivaldi aerial,' *9th Europe Microwave Conference*, pp. 101–105, September 1979.

[63] X. M. Qing and Z. N. Chen, 'Antipodal Vivaldi antenna for UWB applications,' *European Electromagnetics-UWB SP7*, July 2004.

[64] R. W. P. King, *The Theory of Linear Antennas*. 1956.

Index

Antenna
  Base station, 10, 65, 232
  Diamond, 212–14
  Embedded, 171, 180, 227–32, 233
  Measurement, 154–70, 213
  Planar, 5–14, 193–218
  Portable device, 180, 220
  Receive, 195–7, 204–11
  Roll, 183–6
  Square, 5–7, 13, 34, 36, 101, 221
  Transmit, 194–8, 201–11, 212–13, 220
Antenna array, 111–12

Band-gap
  Electromagnetic band-gap, 111, 144
Bandwidth
  Gain, 4, 5, 9, 39, 62, 90, 143, 185
  Impedance, 3, 7, 49–65
  Radiation, 19, 56, 181
Base station antenna, *see* Antenna
Bent
  Planar antenna, 13
Bluetooth, 171
Boresight
  Radiation pattern, 55
Broadband antenna, 39, 179, 204, 206, 209–11

Capacitance, 33, 49, 63, 87, 185, 188, 204
Capacitive loading, 86–92

Chip capacitor, 86–9, 93
Co-pol
  level, 55
Co-to-cross-pol ratio, 5, 9–11, 13, 55, 58, 59,
    72, 90–1, 93, 99, 108
Coplanar design, 22–3, 57–8
Coupling
  Aperture, 23, 85
  Electromagnetic, 56–9, 192
  Feed, 192–3
Cross-pol
  Level, 4, 107
Current distribution, 27, 28, 29, 30,
    72–3, 157

Dipole
  Antenna, 18, 201
Direction
  Radiation, 19, 65
Diversity, 66
Dominant mode, 11, 14, 23, 24, 55, 65, 83,
    97, 103
Dual-band operation, 104, 106, 136, 172

Efficiency
  Radiation, 3, 6, 19, 32, 47, 60, 230
EIRP (Effective Isotropic Radiated Power),
    194, 197

Emission Limit
  Mask, 197–8
E-plane, 8, 9–10, 55, 62, 72–9, 89–93, 95–9, 96,
    108–11, 116, 127, 234, 235

FCC (Federal Communications Commission),
    180, 193, 197, 200, 204
Feed
  Coupling, see Coupling, Feed
  Gap, 10, 63, 150–1, 184, 192, 229
  Probe, 57, 58, 60, 67
  Slot, 14
  Strip, 68, 79, 102, 150
FFT (Fast Fourier Transform)
  Inverse, 212, 213, 214
Fidelity, 195, 204, 209, 210–11
Field
  Distribution, 161, 168, 169, 170
  Electric, 14, 26, 152, 201, 203
  Magnetic, 154
Finite difference time domain (FDTD)
  Method, 156, 195
FR4
  Dielectric substrate, 14, 20, 47, 226, 233

Gaussian
  Pulse, 198–200
GPR (Ground Penetrating Radar), 193
GPS (Global Positioning System), 135, 136,
    172, 197
Ground plane
  Finite-size, 41, 79, 143, 154, 171
  Infinite, 10, 58, 77, 87, 136, 145, 150, 185
  Non-planar, 48, 112, 117
  Planar, 113, 115, 118, 128
GSM (Global System for Mobile
    Communication), 136, 156, 157, 158, 159,
    161, 163

Half-power beamwidth, 9–10, 41, 55, 59, 84,
    108, 125, 234
Handsets
  Antenna, 154–70
  Embedded, 154, 156, 158, 180
H-plane, 9–10, 55, 62, 72–9, 89–93, 95–9, 96,
    108–11, 116, 126, 234, 235

Inverted-F
  Antenna, 135–74

Inverted-L
  Antenna, 135–74
ISM (Industry Scientific Medical)
  Band, 41, 43, 136, 160, 171, 172

Laptop
  Computer, 171–4, 180, 218, 227–32, 233
Linear
  Polarization, 39
L-shaped
  Feed, 56, 92
  Plate antenna, 145
  Probe, 56, 92–100

Matching
  Impedance, 32–6
  Polarization, 196, 197, 218
Matching Factor (MF), 58, 145, 147–8
Medical Imaging
  System, 194
Microstrip line, 39, 135, 155, 168
Microstrip patch antenna, 17–43
Monopole
  Antenna, 181–92
  Planar, 179–236
  Roll, 186–92
Multi-band
  Antenna, 206, 212, 217, 218
  Operation, 144
  Scheme, 198, 199, 206–11, 222
Mutual coupling, 112–16

Omnidirectional
  Radiation pattern, 67, 191, 221

Patch antenna, 17–43
PCB (Printed Circuit Board), 220–7
PCS (personal Communication System), 2, 157,
    158, 161, 163, 168
PDA (Personal Digital Assistant), 135, 136,
    143, 180
Phase
  Linear, 206
PIFA (Planar Inverted-F Antenna), 14, 136, 139,
    142–4, 173
PILA (Planar Inverted-L Antenna), 14, 136,
    144–54
Planar antenna, see Antenna

Plane
E, 8, 9–10, 55, 62, 72–9, 89–93, 95–9, 96,
108–11, 116, 127, 234, 235
H, 9–10, 55, 62, 72–9, 89–93, 95–9, 96,
108–11, 116, 126, 234, 235
Plate antenna
Suspended, 47–130
Polarization
Bandwidth, 2, 4
Circular, 39
Linear, 39
Purity, 5, 10, 22, 55, 62, 149
Probe
L-shaped, 56, 92–100
T-shaped, 57–9
PSD (Power spectral density), 194–5, 202, 205

Q factor, 61

Radiation
Co-pol, 9–10, 55
Cross-pol, 7, 9–10, 55, 58
Gain, 7, 65, 143, 158, 161
Pattern, 158–61
Rayleigh
Pulse, 199–201, 206, 211
Return Loss, 3, 7, 26, 50, 103
Roger4003
Dielectric substrate, 66, 86, 149, 156, 233

S parameter
Measurement, 113, 213, 220, 221
Signal-to-noise (S/N), 198, 206
Single-band
Scheme, 198, 200, 208, 209
Slot
Annular, 49
Capacitive, 49
Center, 81–92
Double L-shaped, 92–100
Edge, 104–5, 108–9
Loop, 49–51
L-shaped, 52
Parasitic, 51
U-shaped, 51
Ω-shaped, 52–6
Small
Antenna, 211, 228
Strip
Feed, 68, 79, 102, 150

Surveillance
System, 194
Suspended pate antenna (SPA)
Array, 66
Capacitive load, 49–51, 86–92
Center-concaved, 75–6
Center-fed, 81–100
Double-L-shaped probes, 92–100
Electromagnetic coupling, 56, 57, 59
Element, 51, 59, 66
Half-wavelength feeding strip, 48, 66
Probe-fed, 66, 69–72, 77–81
Shorting pin, 48, 100, 141
Slot, 49–51, 52–6
Strip-fed, 100
Vertical feeding sheet, 48

Through-wall imaging
System, 194
Transfer function
Amplitude, 206
Radiation, 197, 198, 204, 211
Transmission line, 23, 68–9, 83, 152, 156, 181
T-shaped
Probe, 57–9

U-shaped
Slot, 51
UWB (Ultra-wideband)
Antenna, 195–211
Imaging, 193
Radar, 180
Radio system, 180, 194, 198, 212
Signal, 211
Spectrum, 195
System, 179, 194, 195, 212
Technology, 193–5

Vehicular
Radar system, 194

Wall imaging
System, 193–4
Waveform
Pulse, 204, 206
WLAN (Wireless Local Area Network), 43,
135, 172, 180, 198, 218, 227
WPAN (Wireless Personal Area Network), 228